*INTRODUCTION*
*TO*
*EVOLUTION*

# *INTRODUCTION TO EVOLUTION*

*Fred A. Racle*
**Michigan State University**

**Prentice-Hall, Inc., Englewood Cliffs, N.J. 07632**

*Library of Congress Cataloging in Publication Data*

RACLE, FRED A
Introduction to evolution.

Bibliography: p.
Includes index.
1. Evolution. I. Title.
QH366.2.R26 575 78-25637
ISBN 0-13-482869-0





Printed in the United States of America

10 9 8 7 6 5 4 3 2 1

Prentice-Hall International, Inc., *London*
Prentice-Hall of Australia Pty. Limited, *Sydney*
Prentice-Hall of Canada, Ltd., *Toronto*
Prentice-Hall of India Private Limited, *New Delhi*
Prentice-Hall of Japan, Inc., *Tokyo*
Prentice-Hall of Southeast Asia Pte. Ltd., *Singapore*
Whitehall Books Limited, *Wellington, New Zealand*

*To my wife, Jo Ann*

# Contents

## 9 Phylogenetic Divergence 96



## 10 Human Evolution / One 125

# Preface

Today, the theory of organic evolution is an integral part of the scientific explanation of the natural world. However, despite the general acceptance of this theory by scientists, it remains a poorly understood and even controversial subject to the nonscientist. One reason for this state of affairs is that the subject is more often discussed in terms of its consequences than its processes. Such statements as, "amphibians were the first land veretebrates," or "humans and apes share a common ancestry," do not in themselves provide a clue to the series of careful observations and conclusions that led scientists to their formulation. The purpose of *Introduction to Evolution* is to present an outline of the major observations, and the arguments derived from them, that have led to growth and development of the modern theory of organic evolution. It is hoped that this approach will provide the reader with an understanding of the theory itself and of why it has gained acceptance by the scientist.

The idea for *Introduction to Evolution* grew out of my experiences in teaching undergraduate nonscience majors in the Department of Natural Science, Michigan State University. I am greatly indebted to all members of this department for creating and maintaining an atmosphere in which excellence in teaching is a constant goal. Specifically, thanks are due to Drs. Mohamed O. Abou-el-Seoud, William T. Gillis, Jr., Raymond A. Hollensen, Ralph W. Lewis, Roy H. McFall, George P. Merk, Wilbert E. Wade, and also to Garry T. Voss for their advice and assistance. Special thanks go to Dr. Richard J. Seltin for his thoughtful criticism and friendly encouragement.

# I Introduction

The concept that all living organisms have arisen through the gradual modification of older life-forms dates at least to ancient Greece and was revived as a scientific theory by Buffon, Lamarck, Diderot, and others in the late eighteenth and early nineteenth centuries. However, it was the precision and detail with which Charles Darwin documented the argument in favor of the theory of organic evolution that led to its general acceptance in the closing decades of the last century.

The theory of organic evolution stood in sharp contrast to the beliefs of the Christian Church, which dominated morality, philosophy, law, and science. The position of the Church was *fundamentalist,* maintaining that the word of the Holy Bible is literal truth. The literal interpretation of the Book of Genesis led to belief in a world that: (1) was created by God, (2) was no more than five or six thousand years old, and, (3) was essentially unchanged since the time of the Creation. Change and variability were looked on as exceptions in a universe that was static and immutable.

Religious fundamentalism was reflected in two important tenets of biology: (1) the doctrine of special creation, and (2) the doctrine of fixity of species. Special creation is the belief that every "kind" of organism was specially created by God as part of the overall creation of the universe. The concept of fixity of species is credited to the Swedish botanist Carolus Linnaeus, who established the system of binomial nomenclature. Linnaeus refined the vague Biblical term "kind" into a defined taxonomic category called *species*, and he believed that there had been no change in any species since the time of its creation.

The purpose of this book is to examine the evidence and the arguments by which biologists came to accept the theory of organic evolution. Part of this story is the changing attitude of scientists themselves. Science, like philosophy and religion, attempts explanation of a part of man's experience. Prior to the nineteenth century, biological science was closely allied with Biblical interpretation, and scientists generally viewed their proper role as finding confirmation of the Bible in nature.

A major change in the biology of the nineteenth century was in its *approach* to the study of nature. Whereas religion relies on divine revelation, science shifted to an emphasis on observation as the basis of explanation. Science became a *mechanistic* system, accepting the reality of the physical universe and assuming that an understanding of nature can be based upon observation and interpretation. Being mechanistic, scientists refrain from the use of vital forces or spiritual revelations as a source of explanation. Although this empirical approach is the strength of modern science, it is important to realize that it places limitations on the *scope* of science. Science is incapable of dealing with moral, ethical or spiritual questions because these areas of human experience lack observational evidence upon which explanation can be founded.

Observation alone does not constitute explanation. To explain what has been seen, the scientist must rely on his mental processes to unify his observations through the formulation of *hypotheses* ("educated guesses"). Darwin guessed that the fossils he saw in South America indicated an evolutionary sequence, but he realized that to have any confidence in this hypothesis he would need a great deal of evidence to support his conclusion and so spent thirty years accumulating that evidence.

Once the scientist has sufficient evidence to be confident of his hypotheses, he can formulate a scientific theory. A theory is an *intellectual creation* that attempts to unify a large body of observational evidence and the hypothetical conclusions derived from it. The value of a scientific theory lies in its ability to explain and predict. If a theory is sound, it not only explains the available observations, but it also accommodates new observations and hypotheses formed from them.

In the third century B.C. Democritus postulated a theory that all matter is made up of small indivisible particles called *atoms*. This theory was revised in 1800 by Thomas Dalton as the *atomic theory*. The difference between the two statements is that the atomic theory of Dalton was based on a large body of observational evidence taken from physics and chemistry but that of Democritus was founded on philosophical speculation. Dalton's theory *explained* the behavior of chemical elements and their reactions and the behavior of gases, and was used by Mendeleev to predict the existence of chemical elements undiscovered at the time. The concept of matter being composed of invisible units with definable properties is valuable in explaining many chemical, physical, and biological phenomena.

The point of the foregoing is that a scientist argues from his evidence. If his evidence is sound and his argument logical, he can develop a theory that is both consistent with his observations and useful to science. But, because it is an intellectual creation, a theory cannot be proven true. It can be strengthened by further observations that confirm its predictions, or it can be weakened by new evidence that conflicts with its statements. If a sizable body of negative evidence is accumulated, the theory must either be modified or rejected. Much scientific research is aimed at testing the statements of scientific theories.

Beginning with Chapter 2, we will trace the development of the theory of organic evolution and examine the observations and arguments that have made it the dominant theory of modern biology. In later chapters we will see how the theory is used in the explanation of such diverse events as the great proliferation of the dinosaurs and the origin of man.

# 2 Earth History

## Introduction

A great transition took place in the study of Earth history from the middle of the eighteenth century through the nineteenth century. During this period, scientists sought to explain natural phenomena by emphasizing critical observation and experimentation, rejecting theologically oriented doctrines handed down from the Middle Ages. The increase in sophistication of scientists during this period was marked by the development of such areas of specialized study as stratigraphy, paleontology, comparative anatomy, comparative embryology, and biogeography (the geographic distribution of plants and animals). Knowledge from these fields contributed markedly to the development of the theory of organic evolution proposed by Charles Darwin.

The science of geology grew out of the practical pursuits of mining and engineering, both of which date from the earliest human civilizations. Primitive men studied rock *strata* (layers of sedimentary rock) to determine if they contained valuable minerals such as iron, tin, or gold. Later, stratigraphic studies broadened to include those characteristics that might influence the construction of sewers, roads, and canals. From these pursuits man developed a pragmatic knowledge of rocks and their properties.

As new regions were entered and colonized, the invaders looked for similarities between the rocks of the new regions and the familiar formations they had left behind, an early attempt at geographic correlation of strata. This early interest in rock strata did not include an attempt to explain the origin of the rock layers, but did provide a sizable accumulation of data, which was used by later geologists to formulate theories of Earth history.

Geology emerged as a recognizable field of scientific study during the latter half of the eighteenth century, and the geologists of this period *were* interested in explaining the origin of the Earth and the events responsible for its geologic features. The theories of Earth history formulated by these men reflect both their observational background and the influence of their religious training.

## The Theory of Neptunism

The first generally accepted theory of Earth history was developed during the latter half of the eighteenth century by Johan Gottlob Lehmann and Abraham Werner. This theory proposed that the surface features of the Earth (mountains, plains, oceans, etc.) were formed by a great flood that completely encircled a "nucleus" of primordial matter. As this great flood receded, rock strata were formed from materials present in the flood waters. All surface features of the Earth were assigned to the agency of this one momentous geologic event. Since the time of the great flood, it was thought that there had been little change in the Earth. This theory came to be called *Neptunism,* after the Roman god of the sea.

Neptunist theory gained wide acceptance for several reasons. To begin with, Werner was one of the most influential and respected teachers of the last quarter of the eighteenth century, and his endorsement of Neptunism gave great weight to its statements. Also, Neptunist theory was consistent with the body of observational data in geology. Approximately seventy-five percent of all the rocks of the Earth's surface have been formed from water-lain sediments that have been converted into solid rock. Rocks formed in this way are called *sedimentary rocks* (Figure 2.1). Another point in favor of Neptunist theory was that it was easily accepted by the Church, as the great flood was vaguely reminiscent of the Noachian Flood described in the Bible, and it could be interpreted as representing the single "act of creation" described in the Book of Genesis.

The one recognized exception to Neptunist theory was the formation of *igneous rocks.* Igneous rocks are formed by volcanic activity and were considered to be younger than sedimentary rocks. The younger age of igneous rocks was accepted because: (1) igneous rocks often cut across sedimentary layers, and (2) active volcanoes still exist in the world and erupt

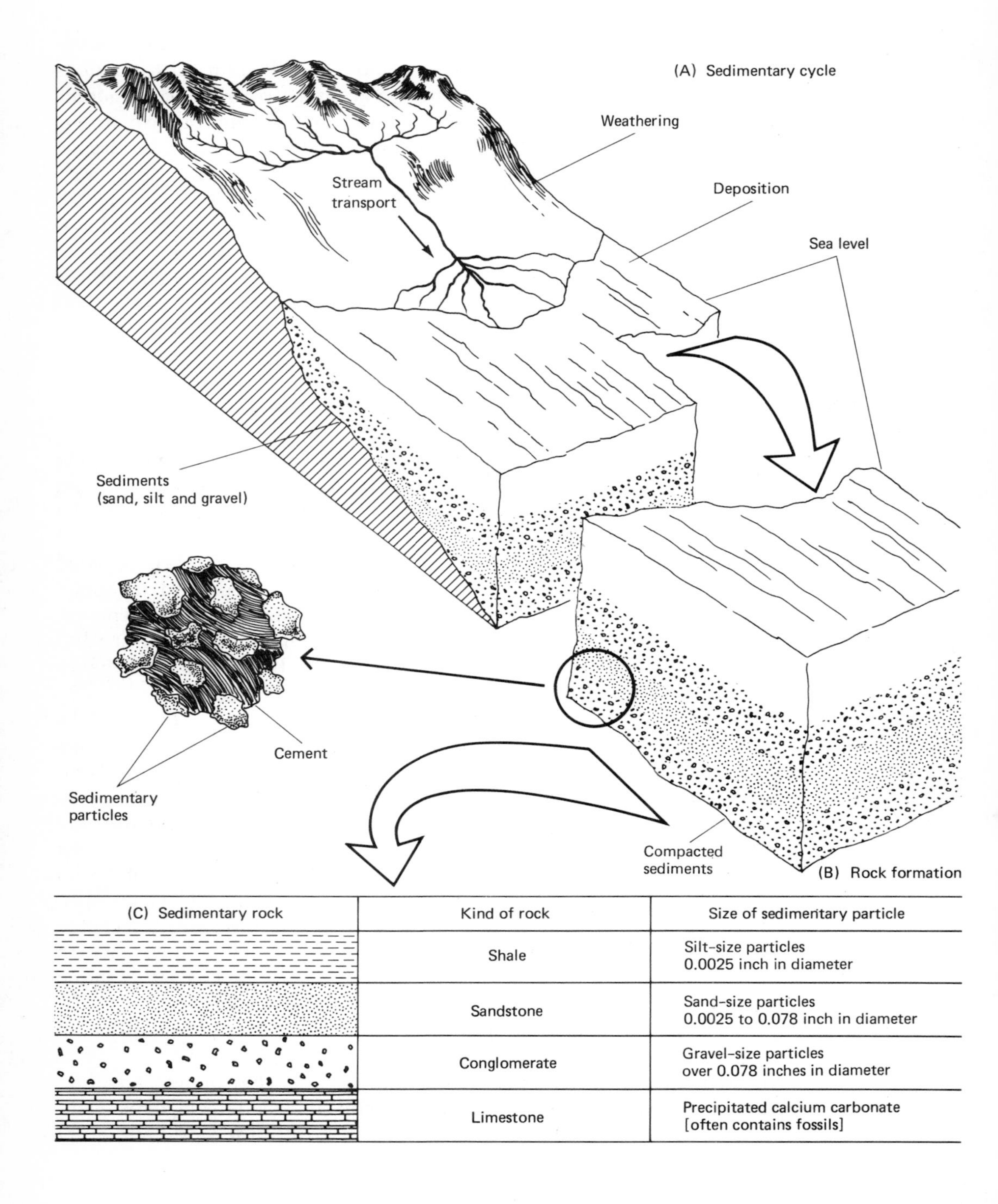


| (C) Sedimentary rock | Kind of rock | Size of sedimentary particle |
|---|---|---|
| | Shale | Silt-size particles<br>0.0025 inch in diameter |
| | Sandstone | Sand-size particles<br>0.0025 to 0.078 inch in diameter |
| | Conglomerate | Gravel-size particles<br>over 0.078 inches in diameter |
| | Limestone | Precipitated calcium carbonate<br>[often contains fossils] |

from time to time to produce lava flows, which cool and harden to form new igneous rocks (Figure 2.2). The younger age of igneous rocks did not present a serious exception to Neptunist theory. Vulcanism was thought to result from the spontaneous combustion of coal deposits within the Earth and was considered a minor event in the generally static, unchanging history of the Earth's surface.

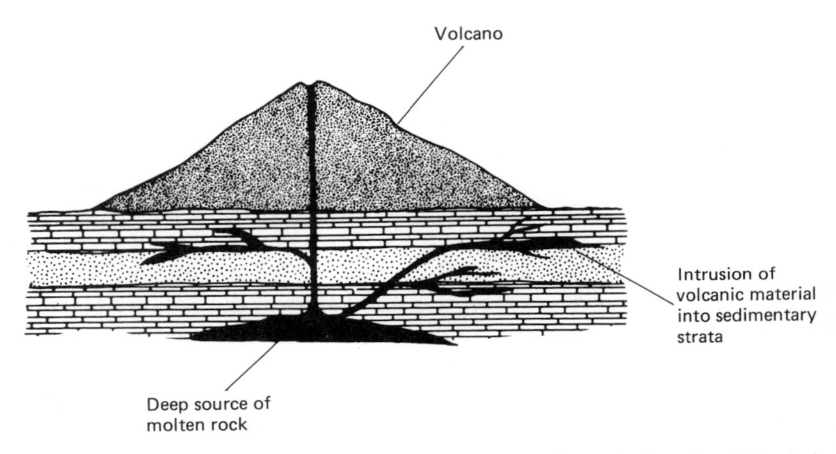


**Figure 2.2** The formation of volcanoes. Volcanoes are formed when deeply buried molten rock moves to the surface along planes of weakness in overlying strata. As part of the process of volcanism molten materials are forced into underground strata (intrusions). Volcanoes are younger than the strata on which they repose and volcanic intrusions are younger than the strata they cut across (crosscutting relationship).

## James Hutton and Uniformitarianism

Toward the end of the eighteenth century, a challenge to Neptunist theory was presented by James Hutton, a gentleman-farmer-physician-geologist. Hutton liked to take long walks through the Scottish countryside—

---

**Figure 2.1** The formation of sedimentary rocks. (A) The sedimentary cycle: 1. the breakdown of rock materials into smaller particles (*weathering*); 2. the *transport* of weathered materials, principally by rivers; 3. the *deposition* of weathered particles at the point where the river enters the still waters of the ocean, forming thick layers of sediments. (B) Rock formation. The thick layers of sediments are compacted by their own weight and cemented together by minerals present in the water (e.g., silica, iron, or calcium). (C) Sedimentary rocks are classified on the basis of the size of the particles of which they are made. Limestone is the exception; it is a precipitate of calcium carbonate.

through the rugged hills, along the sparkling streams, across the coastal plain, and down to the sea. In the course of these outings, Hutton's mind was alive with what he saw and the questions raised by his observations. He studied rock formations, noted the effects of erosion, saw that eroded particles were tumbled along the stream beds, and recognized in the pebbles of the seaside the same type of rock material eroded from the nearby hills.

None of these observations suggested a static Earth to Hutton, nor an Earth that was young and little changed. Instead, he envisioned a dynamic Earth, an Earth that changed day by day, hour by hour. To be sure, the rate of change was slow, but, given enough time, changes of great magnitude could take place on the surface of the Earth.

From his observations and the conclusions he drew from them, Hutton developed his own theory of Earth history. He concluded that the surface of the Earth is continually worn away by erosion, and that the eroded debris is transported by streams until the running waters of the rivers meet the still waters of the oceans. Here, the energy of its transport lost, the eroded rock debris is deposited in layers of sediment that grow very thick and in time are converted into sedimentary rock.

The slow but continuous working of the process of erosion would ultimately result in the reduction of all land surfaces to sea level. That this had not occurred was evidence of a second force at work in the Earth—that of *uplift*. Hutton believed that the sedimentary rock layers formed in the oceans must be gradually elevated to form new land surfaces that replace those being worn away. Hutton found evidence for this part of his theory in the fact that the rocks of the plains and the hills are sedimentary rocks of the type formed from rock debris deposited in the sea. He also noted that the rocks of the highest mountains contain fossils that resemble living organisms found only in the sea. Thus, he concluded, the rocks of the continents must have been formed under water.

Neptunists could agree that the rocks of the Earth were formed under water, and could also agree to the marine (oceanic) origin of the fossils found in these rocks. The disputed point was whether these features had resulted from the recession of the great flood, or from the slow sequence of events suggested by Hutton. Hutton's theory stated that all the Earth's features have been formed through an unending cycle of weathering, transportation, deposition, rock formation, and uplift. This repetitive sequence of events required vast periods of geologic time and certainly violated the concept of a static Earth. More subtly, Hutton's theory presented an entirely new approach to the study of Earth history, interpreting past events from the study of contemporary events and processes. In contrast, Neptunist theory assumed one set of conditions for the origin of the Earth and another set of conditions for all the time since the Creation.

Hutton realized full well that his theory would be unpopular with scientists and theologians. Yet he believed very strongly that the Earth we see today is a product of an endless cycle of erosion and uplift. To Hutton, there was "no vestige of a beginning, no prospect of an end." Hutton's theory came to be called *Uniformitarianism* because of his emphasis on the uniform working of natural forces through long periods of geologic time.

Uniformitarian theory did not gain immediate success, but it did attract a group of adherents who strongly supported Hutton's ideas in the ensuing battle between the Uniformitarians and the Neptunists. Both sides attempted to support their position through the accumulation of empirical data, and this struggle helped swing the emphasis away from academic debate and focus attention on the value of observational evidence in support of a theoretical position.

## The Problem with Fossils

The existence of curious "rocks" bearing resemblance to living organisms has been known for many centuries. Interpretations given to these curiosities of nature were inconsistent and, at times, bizarre. As early as the fifteenth century, Leonardo da Vinci wrote that fossils are the actual representations of previously existing life-forms, but his view was not widely accepted. Throughout the Middle Ages, and even into the 1900s, fossils were commonly looked on as freaks of nature or, more ominously, as works of the devil placed on Earth to confound man and tempt him away from scriptural doctrine.

Gradually, geologists came to accept fossils as being what da Vinci had said they were, the petrified remains of plants and animals that had lived in the past. As such, fossils raised two important questions of interpretation: (1) What was the age of fossil organisms? and (2) Do any of the fossils represent extinct species?

Proof that fossils were of great age, older than the few thousand years allowed by the Neptunists, would provide strong support for the Uniformitarian argument. However, like the rock strata in which they are found, fossils lack observational characteristics that can be used to determine their *absolute age.* Thus, the question of the age of the Earth remained in the realm of theoretical debate.

Initially, the question of extinction presented a similar problem. Proof that some fossils represented species no longer found on Earth would also present a serious challenge to Neptunist theory. However, most fossils familiar to the scientists of the eighteenth century were marine invertebrates (e.g., shellfish, jellyfish, corals) and because the world oceans were as yet unexplored, it was impossible to state with any certainty whether the known fossil species were extinct or *extant* (still alive). This situation changed dramatically with discovery of more fossil types and increased scientific exploration of the continents and oceans during the nineteenth century.

## Cuvier and Catastrophism

Georges Cuvier was professor of anatomy at the Paris Museum of Natural History and a pioneer in the fields of paleontology and comparative anatomy. He was a staunch supporter of Neptunist theory and a strong opponent of the early theories of organic evolution espoused by such noted scientists as Lamarck, Buffon, and Erasmus Darwin (Charles Darwin's grandfather).

Cuvier's research involved the study of fossils found in the sedimentary strata of the Paris Basin region of France, particularly the fossils of land *vertebrates* (animals with a spinal column). There was a general knowledge of the living land vertebrates of the world, and Cuvier's researches led him to the inescapable conclusion that the vertebrate species represented by the fossils of the Paris Basin no longer existed on Earth. Consequently, he was forced to accept the fact that large-scale extinctions of living species had occurred in the past. Since new species had replaced the extinct forms, this conclusion was a direct contradiction of the doctrine of a single creation. Furthermore, the doctrine of fixity of species could not apply to those animals that had become extinct.

Cuvier also noted that while some strata of the Paris Basin contained fossils of marine origin, others contained fossils of freshwater species. Since it was not conceivable that a single flood had been composed of both fresh and salt water, Cuvier concluded that strict adherence to Neptunist theory was no longer possible.

One alternative to the doctrines of special creation and fixity of species was organic evolution—the theory that modern species had evolved from the extinct species fossilized in the sedimentary rock strata. This alternative was not acceptable to most scientists of the early nineteenth century. Instead, the English geologists Jameson and Buckland postulated that there had not been a single great flood but an alternating series of floods and sudden elevations of the Earth, each event of short duration but of cataclysmic proportions. These cataclysms of nature resulted in the local extinction of many species of organisms, with life being replenished by a series of creations occurring at the end of each upheaval.

This theory, called *Mosaic Catastrophism*, gained some measure of acceptance among theologians because they could equate the last flood with the biblical account of Noah's flood, and the series of creations was looked on as an "evolution in thought" on the part of the Creator. The theory also retained the required short duration of Earth history and maintained that the Earth had been little changed since the last flood.

## The Ascendancy of Uniformitarianism

Clearly, there was a great deal of uncertainty in the geological sciences during the first years of the nineteenth century. A growing body of observational evidence had weakened the older theory of Neptunism, but neither the Uniformitarian theory of Hutton nor the theory of Mosaic Catastrophism could gain ascendancy. Finally, Charles Lyell, a follower of James Hutton's Uniformitarian theory, set about writing a book that he hoped would unify geologic thought and formalize geologic explanation under the guidance of Uniformitarian theory. The result of Lyell's efforts was the three-volume *Principles of Geology,* the first volume of which was published in 1830. This book presented a clear and persuasive argument for the value of the Uniformitarian approach in organizing geologic explanations. Thereafter, geologists either had to challenge Lyell's Uniformitarian explanations or to accept them, at least tacitly.

## The Geologic Time Scale

With the growing acceptance of Uniformitarian theory, geologists turned their attention to the related problems of organizing their knowledge of the various rock strata of the world, and to attempts to determine the absolute age of the Earth. As mentioned above, there are no observable attributes of rocks or fossils that provide any clue as to their actual age. However, there are methods by which geologists can determine the *relative age* of rock layers.

The *principle of superposition* was clearly stated as early as the seventeenth century (Figure 2.3). Because sedimentary rocks are formed from debris deposited by water, it was recognized that the bottom-most layer of sediments was deposited first. Therefore, the bottom layer in any undisturbed sequence of sedimentary rocks is the oldest rock of that series; the uppermost layer the youngest. Superposition provides a relative age for the rocks of any single sequence but tells us nothing about their absolute age.

Another principle used to determine relative ages of mixtures of sedimentary and igneous rocks is the principle of *cross-cutting relationships* (Fig-

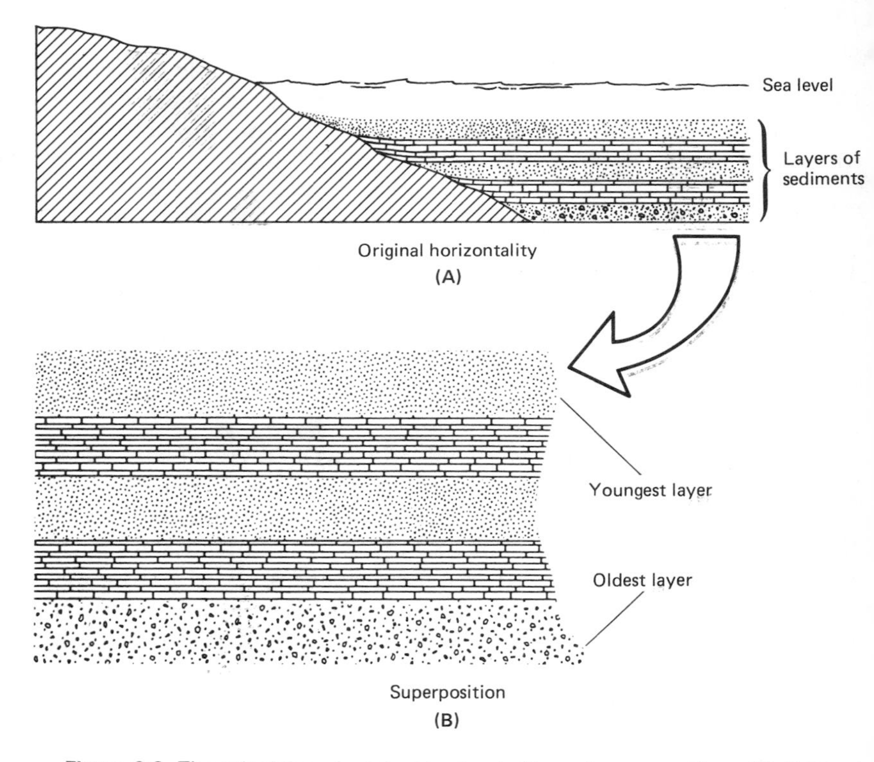


**Figure 2.3** The principles of original horizontality and superposition. (A) Original horizontality. As sediments accumulate in the sea, the natural sorting action of the water distributes them in horizontal layers. When these layers are compacted and cemented into sedimentary rock, the layers retain their horizontality. (B) Superposition. When sedimentary rock layers are elevated above sea level they retain the same relative sequence they had when formed. In such sequences of sedimentary strata, the bottommost layer is always the oldest and the uppermost layer the youngest.

ure 2.2). Like superposition, this principle tells us nothing about the absolute ages of the rocks involved.

While these principles of rock dating proved valuable to geologists in their interpretation of local rock formations, they were of no value in comparing the ages of strata from different geographic regions. There are no unique properties of any rock stratum that permit geologists to be certain that a sandstone or shale layer found in France is the same age as similar layers found in England or Germany. For this reason, the strata of each geographic region were treated as distinct units.

The solution to the problem of correlating the relative ages of rock strata from different regions was provided by a geologist–engineer named

William Smith. Smith travelled throughout England working on the construction of roads and canals, and through his wide experience in the field found that he could compare sedimentary strata of different regions on the basis of the fossils they contained. Smith concluded that strata of a specific age (as determined by superposition) contain a typical group of fossils, called the *fossil assemblage* (Figure 2.4). That particular fossil

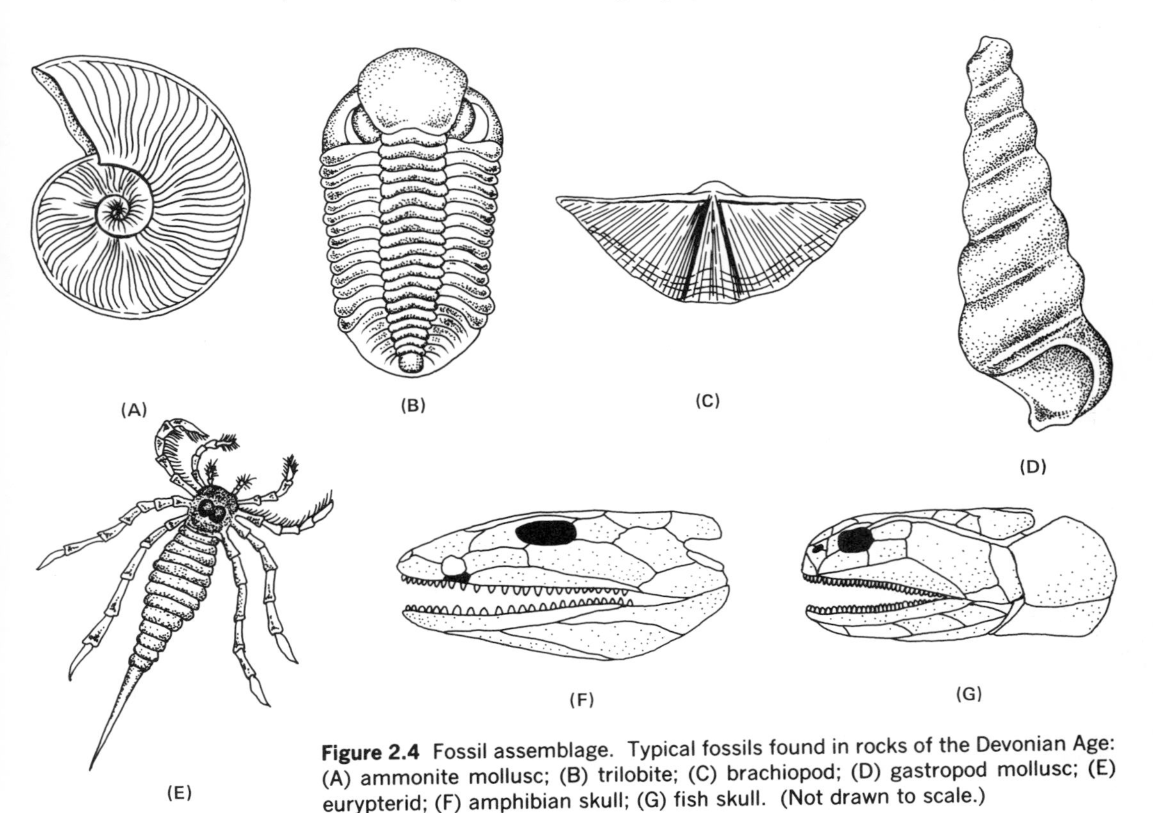


**Figure 2.4** Fossil assemblage. Typical fossils found in rocks of the Devonian Age: (A) ammonite mollusc; (B) trilobite; (C) brachiopod; (D) gastropod mollusc; (E) eurypterid; (F) amphibian skull; (G) fish skull. (Not drawn to scale.)

assemblage would not be found in rocks of any other age because those fossils represented plants and animals that had lived, died, and were entombed only during the time period when this stratum was being formed. Other geologists soon acknowledged Smith's conclusion that fossil assemblages could be used to correlate the ages of the rock strata in which they were found, and they formalized the principle into the law of *organismic succession.* The use of organismic succession to determine the age of rock strata from different geographic regions is called the principle of *fossil correlation* (Figure 2.5).

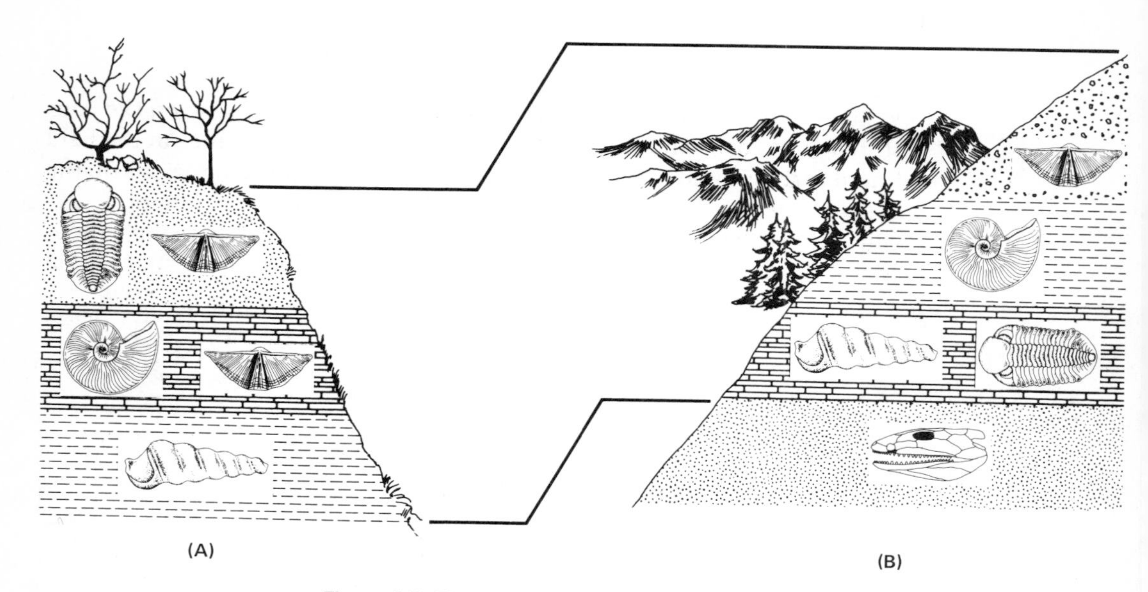


**Figure 2.5** The principle of fossil correlation. Rock strata from two geographic locations (A and B) contain similar fossils and are therefore considered to have been formed during the same geologic period.

Combining the use of the principles of superposition and fossil correlation, geologists were able to make valid time comparisons between rock strata from different geographic locations. The relative age of the rock layers within any sequence could be determined by superposition (Figure 2.6). The typical fossil assemblage of these strata could be noted and compared to the fossil assemblages of rock strata from other regions. By repeating this procedure over and over, within and between continents, the relative ages of rock strata from all over the world could be determined, and a composite stratigraphic column could be developed. From interpretation of the characteristics of rock strata and the fossils they contain, scientists have been able to reconstruct a broad outline of the geological and biological events of Earth history. This generalized sequence of events has been organized into the Geologic Time Scale (see inside back page). The Geologic Time Scale is the geologist's interpretation

---

**Figure 2.6** Development of the time–stratigraphic column. (A) By superposition it is known that $A_1$ is the oldest layer in rock series *A*. (B) By superposition, $B_1$ is the oldest layer in rock series *B*. Because layers $A_3$ and $B_1$ have similar fossils, they are assumed to be the same age (fossil correlation), and a relative time sequence between rock series *A* and *B* has been established. $A_1$ is the oldest rock layer and $B_3$ is the youngest of the two series. (C) Rock layers $C_1$ and $B_3$ are the same relative age (fossil correlation) and layers $C_2$ and $C_3$ are the youngest of all layers. (D) Composite time–stratigraphic column assembled from rock series *A*, *B*, and *C*.

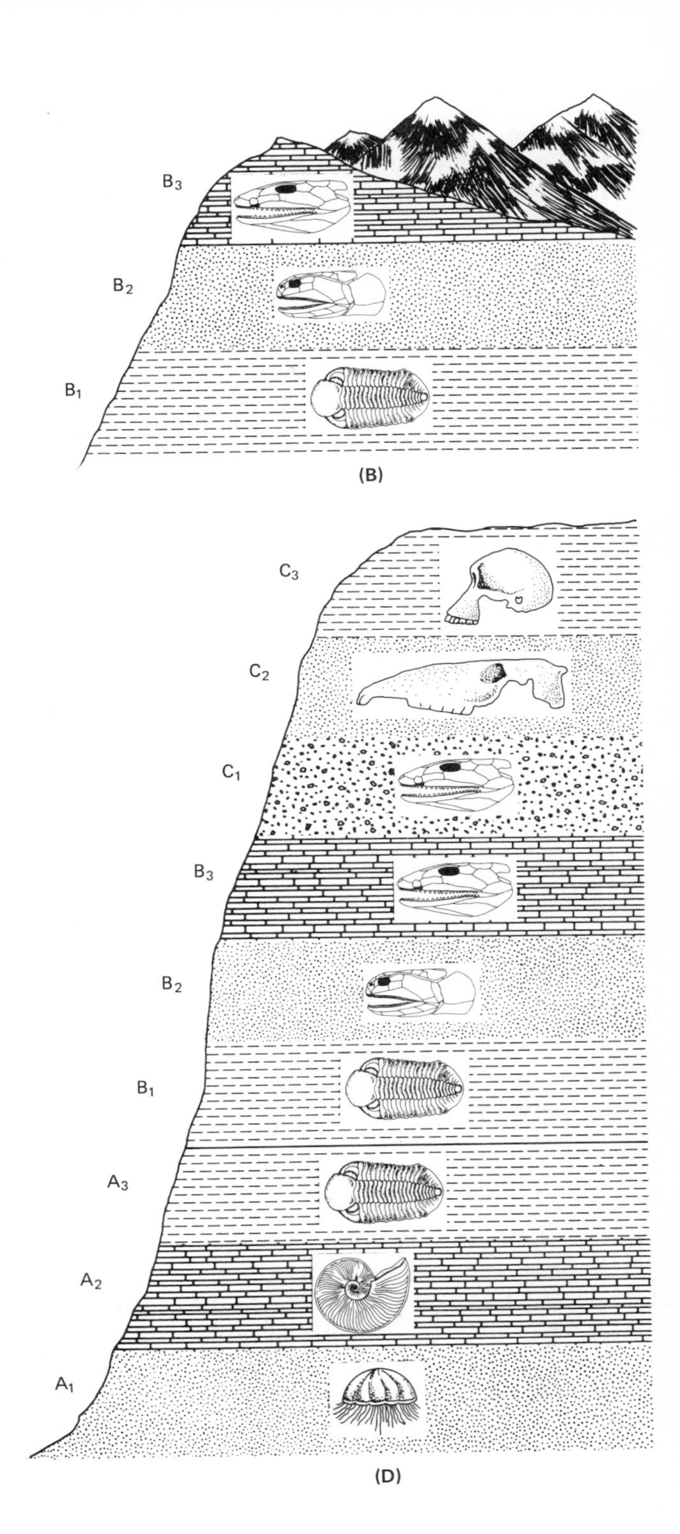


(B)

(D)

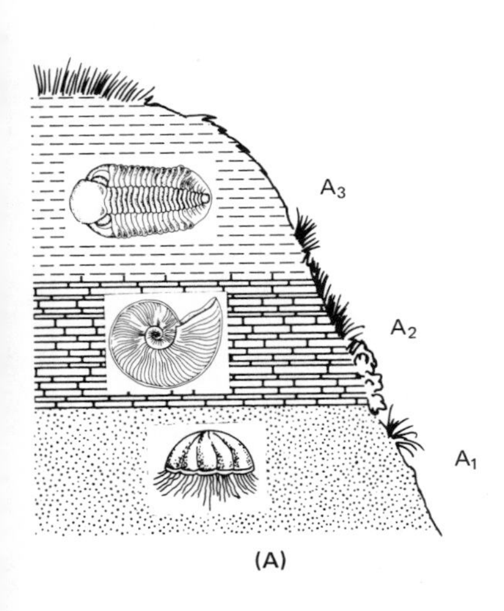


(A)

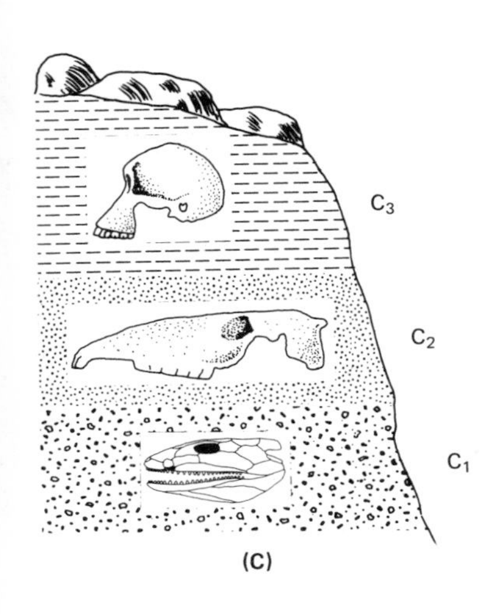


(C)

of Earth history, based on observational data and interpretive principles derived from his observations. The time units of the Geologic Time Scale were originally derived from inferred rates of organic evolution and, more recently, from radiometric dating of rock strata.

The Geologic Time Scale is divided into major units that reflect periods of distinct geological and biological activity. These major divisions are called *geologic eras.* The modern names for these eras are: Precambrian Era, Paleozoic Era, Mesozoic Era, and Cenozoic Era. Each era is subdivided into geologic periods, such as the Triassic, Jurassic, and the Cretaceous Periods of the Mesozoic Era. Geologic periods are units of recognizable geological and biological activity within the larger divisions. Finally, the geologic periods of the Cenozoic Era have been further subdivided into units called *epochs* (Table 2.1). Discussion of specific events of the geologic eras, periods, and epochs is presented in Chapters 8, 9, and 10.

**TABLE 2.1 Divisions of Geologic Time**

| MILLIONS OF YEARS AGO | ERA | PERIOD | EPOCH |
|---|---|---|---|
| 0.01 | Cenozoic | Quaternary | Recent |
| 2 | | | |
| | | | Pleistocene |
| 5 | | Tertiary | |
| | | | Pliocene |
| 23 | | | |
| | | | Miocene |
| 38 | | | |
| | | | Oligocene |
| 54 | | | |
| | | | Eocene |
| 65 | | | |
| | | | Paleocene |

# The Age of the Earth

The Geologic Time Scale provides a strong organizing principle for the study of Earth history. Before development of the Geologic Time Scale, it was only possible to construct unrelated, local histories of isolated geographic areas, but use of the fossil record and the principle of fossil correlation permitted reconstruction of an overall history of life.

Still, the problem of assigning an absolute age to rocks and fossils persisted. Geologists had no way of knowing if all the events of Earth history had occurred within a few thousand years or had taken place over the millions of years predicted by Uniformitarian theory. By the last third of the nineteenth century, the age of the Earth was the central issue between those who supported the newer scientific theories of Uniformitarianism and organic evolution and those who retained a belief in the older theologically oriented theories of the eighteenth century. Thus, it became imperative that a method be found to date accurately the rocks of the Earth.

It was theorized that the absolute age of rock layers could be determined if a geologic process was found that: (1) had taken place since the origin of the Earth, (2) had occurred at a constant rate throughout geologic time, and (3) yielded a collectible, measurable product. Scientists of the latter part of the nineteenth century sought such a process without success. Estimates derived from the rate of formation of sedimentary strata or the rate of accumulation of salt in the sea were based on such tenuous assumptions and incomplete knowledge that they proved to be wildly inaccurate.

In the first decades of the twentieth century it was discovered that the phenomenon of radioactivity could be used to determine the absolute age of rocks. An understanding of the principle of *radiometric dating* can best be approached through a brief discussion of the nature of radioactivity itself.

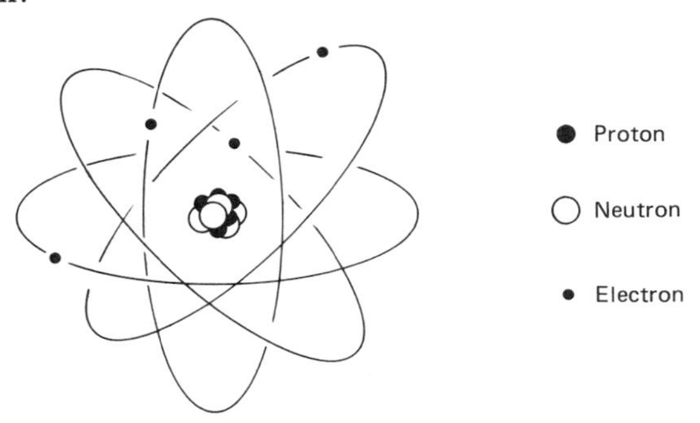


**Figure 2.7** The proton–neutron model of the atom.

The atom is made up of three parts—the proton, the neutron, and the electron (Figure 2.7). The proton and the neutron appear in the nucleus of the atom, whereas the electron orbits the atomic nucleus, somewhat in the way planets circle the sun. Atoms of different elements differ in the number of protons, neutrons, and electrons present. For example, the hydrogen atom contains one proton and one electron; the helium atom two protons, two neutrons, and two electrons; the oxygen atom eight protons, eight neutrons and eight electrons; and the uranium atom 92 protons, 146 neutrons, and 92 electrons. The chemical symbol of an atom is often written with the total number of protons plus neutrons appended (e.g., hydrogen as H-1; helium as He-4; oxygen as O-16; and uranium as U-238). In most atoms the nucleus is very stable, but in others the nucleus has an unstable ratio of protons to neutrons. These unstable nuclei are radioactive and emit nuclear radiations until a stable ratio of protons to neutrons is achieved. In the process of giving off radioactive-decay products, the nucleus itself is changed—often to the extent that a new element is formed. For example, as a result of radioactive decay, uranium-238 is converted into lead-206.

An important characteristic of radioactivity is that not all radioactive nuclei break down at the same time. If we begin with 5000 atoms of the radioactive isotope strontium-90, we will find that one-half of the radioactive nuclei will disintegrate in a thirty-year period, leaving 2500 nuclei yet to decay. After another thirty years, one-half of the remaining nuclei will decay, leaving 1250 radioactive nuclei. This pattern will continue until all radioactive nuclei are gone. Physicists cannot explain why all radioactive nuclei do not disintegrate at the same time, but they do know that every radioactive element has its own specific *half-life* (i.e., the time required for one-half of all radioactive nuclei to break down). The half-life of uranium-238 decaying to form lead-206 is 5510 million years. The half-life of potassium-40 to argon-40 is 1300 million years.

The half-life of a radioactive isotope is exceptionally constant and is not altered by common physical or chemical means. Furthermore, it is generally agreed that radioactive materials such as uranium-238 and potassium-40 were present at the time the Earth was formed. Therefore, the radioactive decay of an element such as uranium-238 meets all the requirements for a process that can be used to calculate the absolute age of the Earth: (1) it is a process that has taken place since the origin of the Earth, (2) it has occurred at a constant rate, and (3) it produces a stable, measurable product (i.e., lead-206).

The application of these principles to the dating of rock strata can be illustrated by the following hypothetical example. Let us assume that a particular rock contains the radioactive element oldium. Oldium has a half-life of twenty-four hours. If we start at noon today with thirty-two atoms of oldium, at noon tomorrow sixteen atoms of newlium will have been formed (Figure 2.8B). In another twenty-four hours one-half of the *remaining* oldium atoms will have changed to newlium so that the ratio of oldium to newlium now stands at 8:24. This sequence of radioactive decay is completed in the table in Figure 2.8C. Once a table such as this has been developed, it can be used to determine the age of rock samples that contain the radioactive element. For example, if our rock sample contains a ratio of one oldium atom to 31 atoms of newlium, the sample is five days old.

The radioactive materials most commonly used in determining the age of rock strata are uranium-238 and potassium-40. Radiometric dating has confirmed the great antiquity of the Earth predicted by James Hutton and Charles Darwin: the age of the Earth as determined by radiometric dating appears to be 4½ billion years.

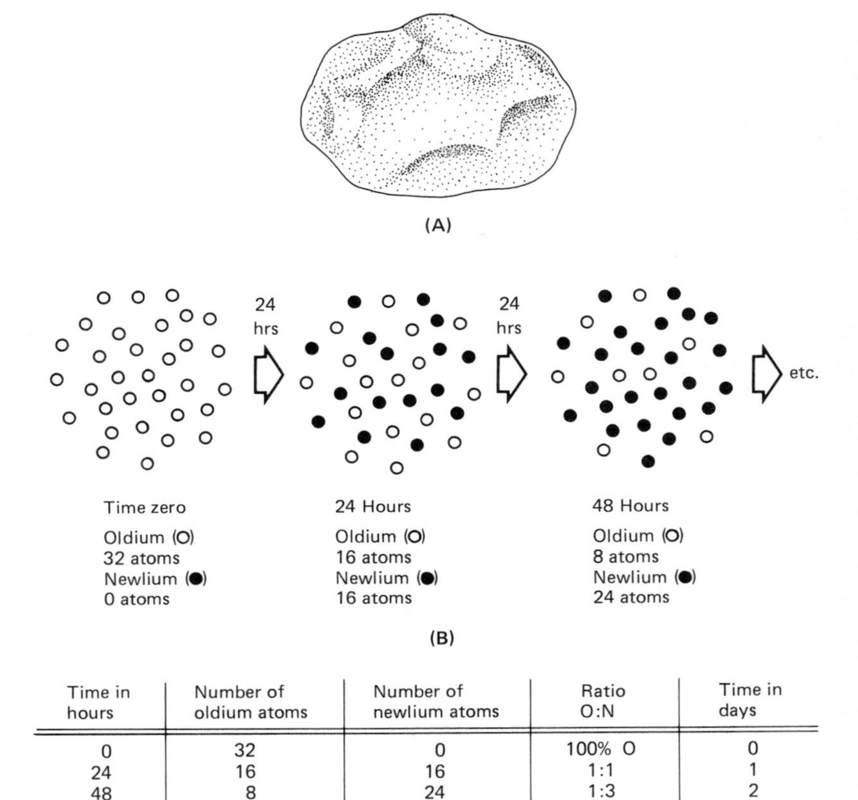


| Time in hours | Number of oldium atoms | Number of newlium atoms | Ratio O:N | Time in days |
|---|---|---|---|---|
| 0 | 32 | 0 | 100% O | 0 |
| 24 | 16 | 16 | 1:1 | 1 |
| 48 | 8 | 24 | 1:3 | 2 |
| 72 | 4 | 28 | 1:7 | 3 |
| 96 | 2 | 30 | 1:15 | 4 |
| 120 | 1 | 31 | 1:31 | 5 |
| 144 | 0 | 32 | 100% N | 6 |

(C)

**Figure 2.8** The use of radioactive decay in dating rock samples. (A) The original rock sample contains 32 atoms of the radioactive element oldium and no newlium. (B) The breakdown of oldium nuclei to form newlium atoms is depicted schematically. The open circles represent oldium atoms, the shaded circles represent newlium atoms. The half-life of decay for oldium to newlium is 24 hours. (C) Tabular data showing the number of oldium versus newlium atoms at 24-hour intervals. The age of this rock sample can be determined from measurement of the ratio of oldium to newlium atoms present in the rock sample.

Modern religious fundamentalists do not accept the validity of radiometric dating, retaining the belief that the age of the Earth is no more than a few thousand years. However, radiometric dating is accepted and widely used in the physical and biological sciences. It is significant that its dates coincide with independent estimates derived from the theoretical considerations of astronomers.

# 3 Charles Darwin and the Theory of Organic Evolution

## Introduction

Charles Darwin was born in 1809, and coming from a well-to-do family, he received a classical education that his father hoped would lead to a medical degree from Edinburgh University. But Darwin was not interested in medicine and found the practice of surgery, in the time before anesthesia, too horrible to countenance. He turned from medicine to preparation for the clergy, but found that the healing of souls held no more fascination for him than the healing of bodies. Thus, at the age of twenty-two, Charles Darwin embarked on the sailing ship H.M.S. Beagle as ship's naturalist—an undertaking much more in tune with his personal interests and aspirations (Figure 3.1). Despite his generally uninspired record in formal education, Darwin was an intelligent young man intensely interested in the study of nature. While attending school, he had associated himself with some of the leading geologists and biologists of the time and had become a trained scientist and critical observer of natural phenomena.

At the time he sailed from England on the Beagle (a survey ship), Darwin was not an evolutionist. His training was in the traditional beliefs of the Anglican Church and the science of the period, and he generally

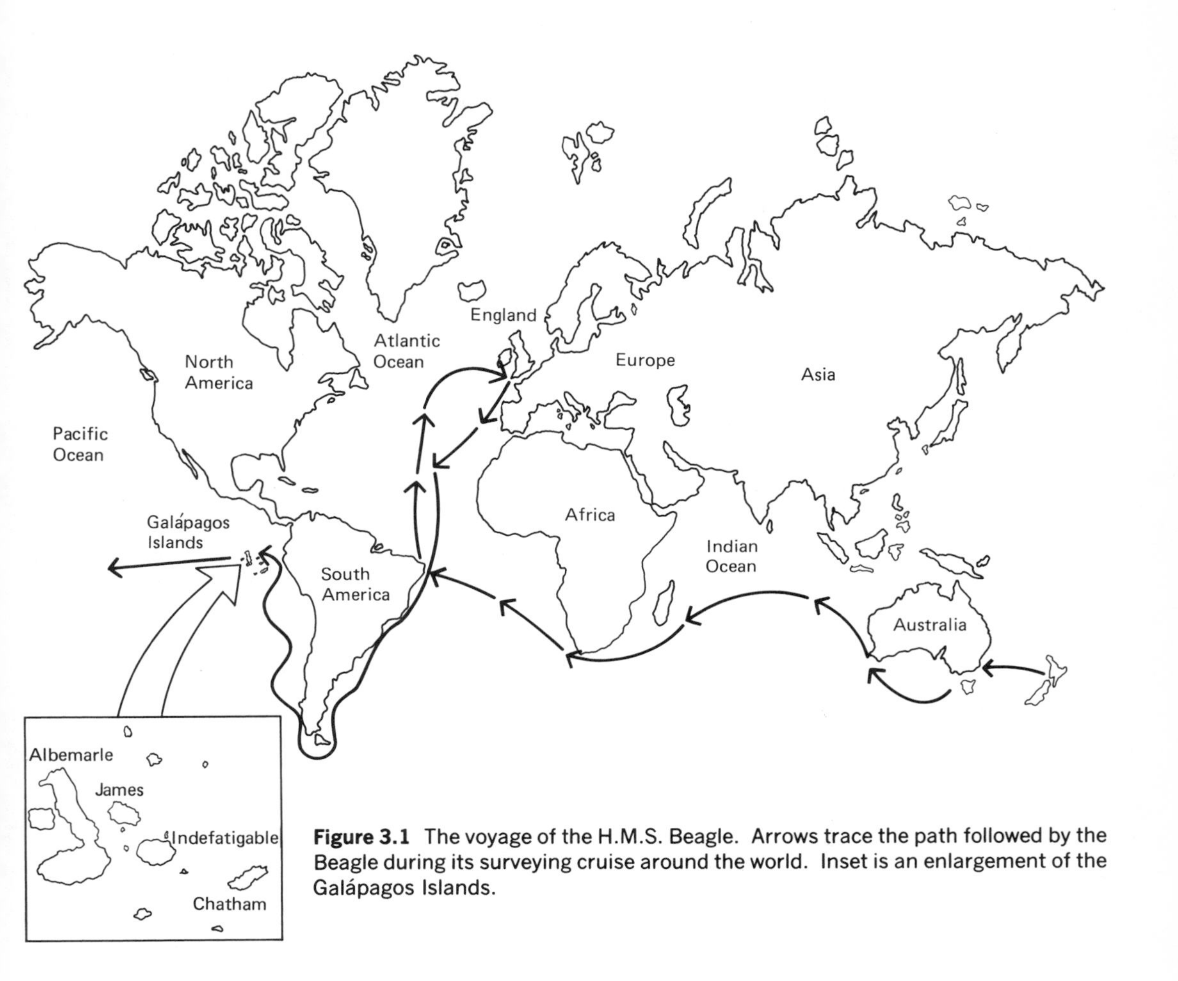


**Figure 3.1** The voyage of the H.M.S. Beagle. Arrows trace the path followed by the Beagle during its surveying cruise around the world. Inset is an enlargement of the Galápagos Islands.

accepted these beliefs. Yet the observations he was to make during his five-year absence from England were to alter greatly his preconceptions and lead him to a period of intellectual activity that culminated in the publication of *On the Origin of Species* in 1859.

During the long voyage of the Beagle, Darwin saw a wider range of living organisms in their natural habitats than had any trained observer before him. These observations greatly influenced his thinking about nature, and in 1837 he wrote that he had been " . . . greatly struck . . . on (the) character of South American fossils and species on the Galápagos Archipelago," and added that these observations were the origin of all his views concerning organic evolution.

Before his departure from England, Darwin was given a copy of the first volume of Lyell's *Principles of Geology*. Darwin was an enthusiastic geologist and Lyell's argument for the slow working of geologic forces over vast periods of time provided a philosophical background against which Darwin could evaluate the observations he was to make over the next five years.

The observations that Darwin made during the voyage can be divided, for convenience, into three categories: (1) those involving the geographic distribution of species; (2) those involving the fossil assemblage of South America; and (3) those involving the fauna and flora of the Galápagos Islands.

## The Geographic Distribution of Species

As he traveled around the world on the Beagle, Darwin noted that the fauna and flora he encountered changed with both latitude and altitude (Figure 3.2). As the Beagle worked its way along the coastline of South America, he observed that the species that inhabited the tropical regions

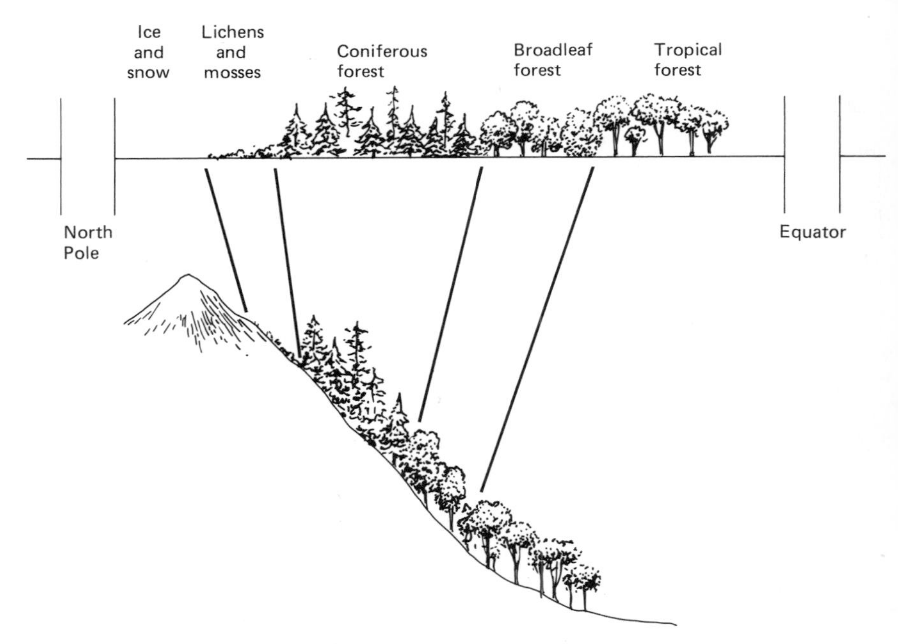


**Figure 3.2** Similarity in vegetation patterns with change in latitude and altitude. An overall similarity in vegetative pattern is found as one proceeds northward from the equator, or from near sea level to the summit of a high mountain because similar temperature and rainfall gradients are experienced with change in altitude and change in latitude.

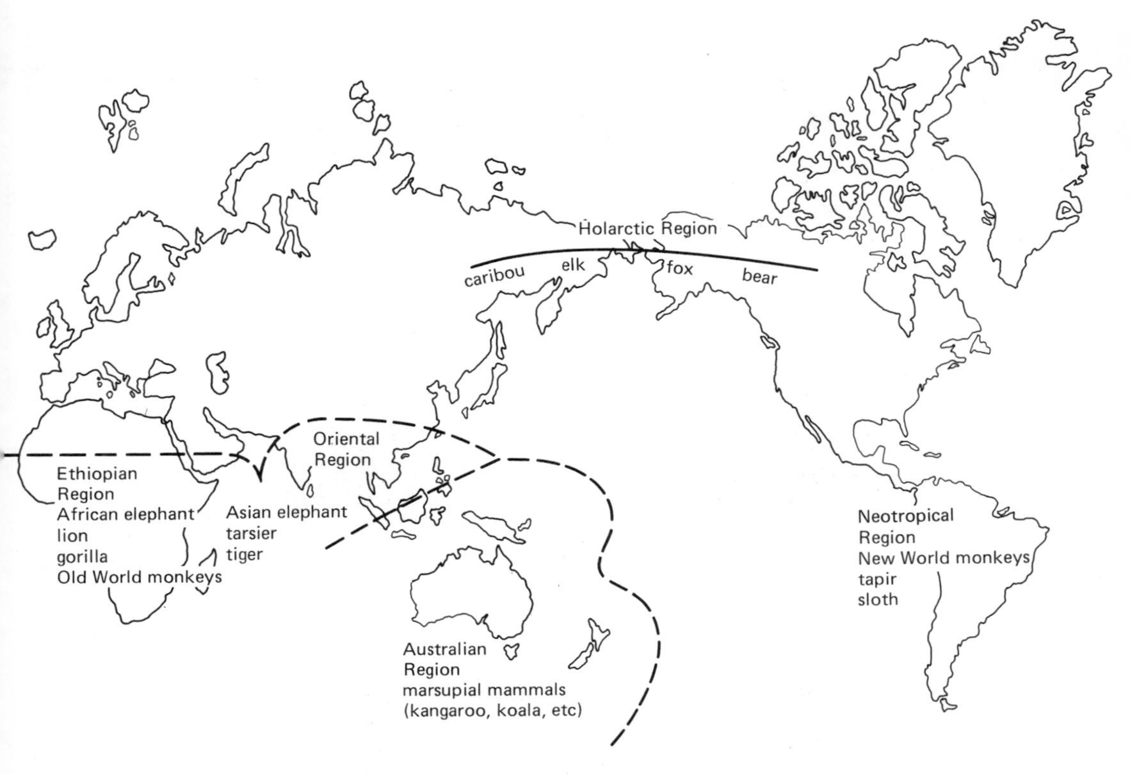


**Figure 3.3** Biogeographical regions of the world. Each region is isolated from the others by geographic barriers such as oceans, mountains, or deserts. Each region has its own characteristic flora and fauna.

of Brazil were different from those found in the frigid wastes of Tierra del Fuego, and he observed that a similar pattern of change occurred as one climbed a high mountain. These observations raised the question in Darwin's mind of *why* separate and distinct species had been created for each geographic region.

Even more difficult for Darwin to explain was the fact that one does not find the same species in similar habitats in different parts of the world. For example, the deserts of South America contain a different selection of plants and animals than the deserts of Australia. The same is true of the woodland creatures typical of forests of different geographic regions. It seemed to Darwin that these phenomena contradicted the basic intention of the doctrine of special creation. He thought a more likely explanation was that the environment had molded the organisms to the rigors of each particular habitat. He gradually discerned that geographic

**Figure 3.4** Convergent adaptations in marsupial and placental mammals. Mammals of diverse origin that have adapted to similar ecological niches. Pictured above are (A) the Australian marsupial *Dasyuris* and (B) the American short-tailed weasel of the genus *Mustela*. Both animals are small terrestrial carnivores. (C) The "sugar glider" *Petaurus breviceps* is the Australian marsupial equivalent of (D) the American placental "flying squirrel" *Glaucomys sp.* Both animals are nocturnal tree dwellers. This is an example of an evolutionary trend known as *convergent evolution.*

barriers, such as mountains or oceans, had separated the developing species for long periods of time and concluded that this *geographic isolation* was responsible for the development of new and distinct species.

Darwin's observations concerning the geographic distribution of species are supported by subsequent investigation. Today, six major biogeographic zones are recognized, each of which contains a typical grouping of animal species (Figure 3.3). A striking example of geographic influence on species distribution is the almost complete limitation of marsupial mammals to the Australian regions, where they occupy all the habitats typically inhabited by placental mammals in other biogeographical zones. Figure 3.4 shows that Australian marsupial and North American placental mammals that occupy comparable environmental niches exhibit close similarities in adaptive traits.

## The Fossil Assemblage

Darwin was especially impressed by the fossil assemblage found in South America. Accepting fossils as representations of extinct organisms, Darwin noted that many species of South American fossils are no longer found on that continent. He interpreted the fossil assemblage as representative of an uninterrupted sequence of change from the oldest fossil form to the youngest. Thus the fossil record provided evidence of a gradual

**Figure 3.5** Extinct and extant South American mammals. The armadillo, guanaco, and tree sloth that inhabit South America today show close resemblances to the extinct glyptodon, camelops, and ground sloth whose fossilized remains are also found in South America.

modification of traits in which the most recent extinct forms bear the strongest resemblance to species found in South America today. This idea of progressive change contrasted strongly with the theories of fixity of species and Mosaic Catastrophism.

The type of observation and interpretation made by Darwin in South America has since been repeated around the world. The fossil assemblage of each geographic region varies, but within each region there occurs a continuous *progression* in fossil forms, with the recent fossil species most closely resembling the living species of that region (Figure 3.5).

## The Galápagos Islands

The Galápagos Islands have achieved a fame from Darwin's visit to their shores that far transcends their physical charm. The Galápagos are volcanic islands lying some 600 miles west of Ecuador, and, being of volcanic origin, are younger than the continental land mass of South America.

According to modern interpretation, the present Galápagos Archipelago is the latest product of a sequence of volcanic eruptions that began over 40 million years ago near the coast of Ecuador. The older volcanic islands arose, existed for a time, and have since subsided. Those first islands, near the shoreline of South America, were populated by organisms that were rafted to the islands on floating logs or debris, or were transported to the islands on wind currents. As newer islands formed, this pattern of dispersal was repeated, and as older islands receded below sea level, the younger islands were left isolated, far from land.

The environmental conditions of the new land surfaces were much harsher than those of the mainland, and a strong element of chance was involved in the survival of organisms transported to them. Few individuals reached the islands and even fewer were able to become established. But those organisms that found conditions suitable for their survival flourished because of the lack of competition for resources and the general lack of predators.

When Darwin traveled through the Galápagos Islands in the 1830s, he noted unmistakable differences within single species of organisms from island to island. An example of inter-island variation within a species is provided by the giant tortoises that gave the islands their name (*galápagos* means tortoise in Spanish) (Figure 3.6). Darwin reasoned that the great variety of tortoises found in the islands must be descendants of relatively few parent animals of a single species that had reached the islands in the past. His inescapable conclusion was that geographic isolation had led to the development of morphological differences as individual tortoise populations gradually adapted to the variations in environmental conditions that existed on the different islands.

**Figure 3.6** Inter-island variation in neck length of Galápagos tortoises. (A) Short-necked turtle of Albemarle Island. (B) Long-necked turtle of Indefatigable Island. The tortoises of the drier Indefatigable Island have evolved a long neck that allows them to reach the sparse vegetation found on this island.

## Darwin's Finches

Probably the best-known observation made by Darwin during the voyage of the Beagle concerns the finch population of the Galápagos. These birds have often been used to illustrate the effect of geographic isolation on the development of variation within species. Finches are common birds that enjoy a wide geographic distribution. In general, they are seed eaters, and their short, stout beak provides an efficient mechanism for this type of feeding. However, in the Galápagos Islands, Darwin discovered a surprisingly wide variation in the size and shape of finch beaks. In addition to the typical stout-beaked seed eaters, Darwin found finches with beaks highly modified for feeding on a wide variety of other foods (Figure 3.7). Darwin reasoned that the first finches to reach the islands were able to increase rapidly in number because of the lack of competition for food and the absence of predators. The increasingly larger finch population soon outstripped the supply of seeds, thus causing some individuals to seek out

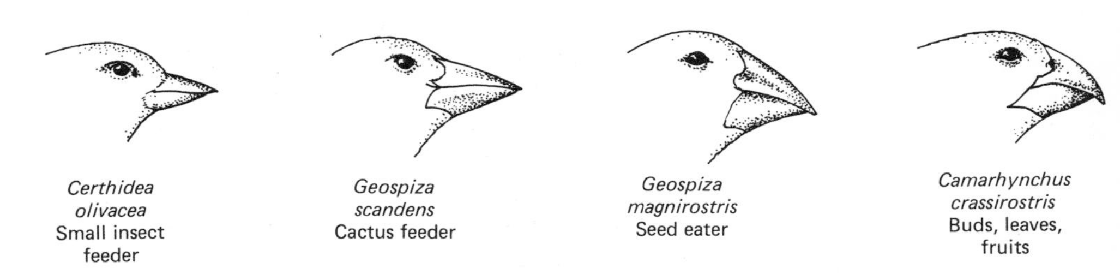


**Figure 3.7** Variation in finch beak as related to food type.

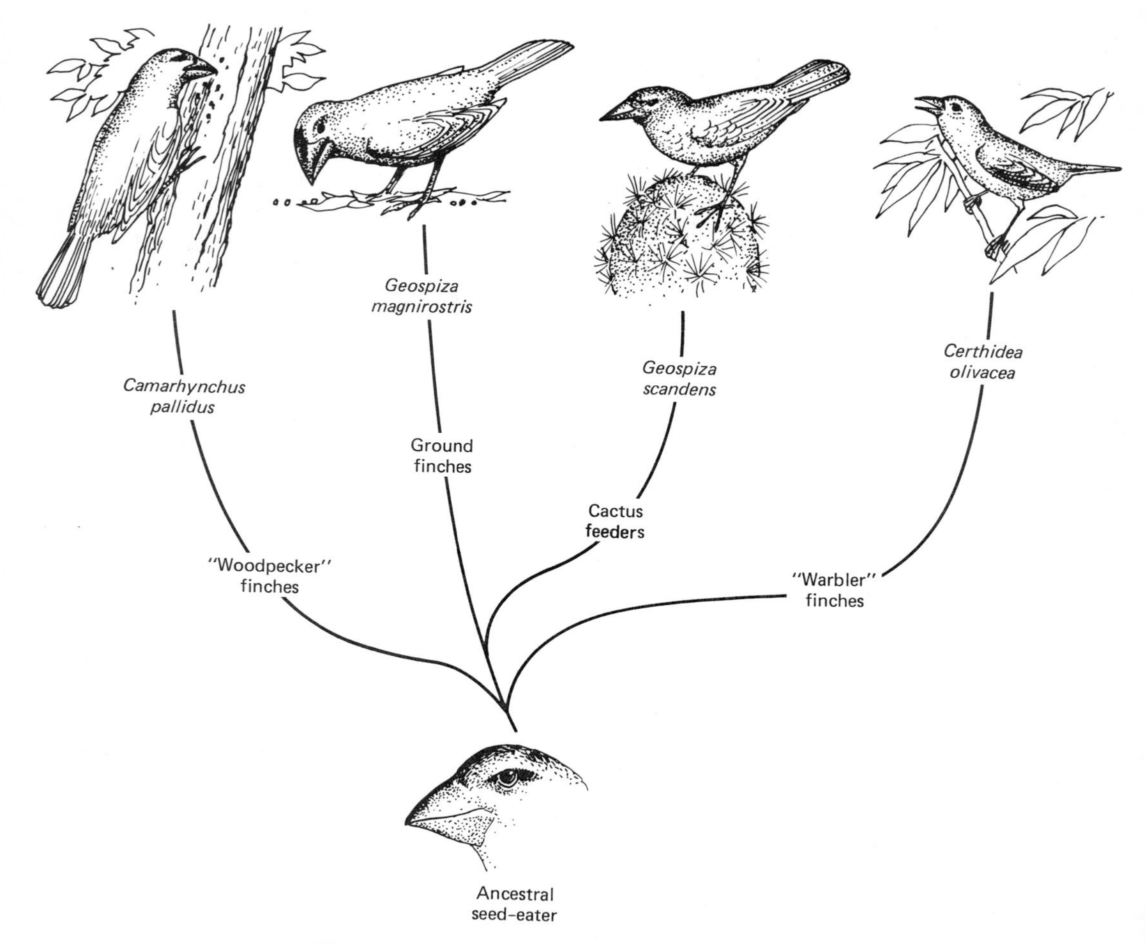


**Figure 3.8** Adaptive radiation of finches.

alternate food sources (e.g., insects, leaves, or fruit). Finding less competition for these foods, subpopulations developed that continued to feed on the less restricted resources. Within each subpopulation, individuals appeared with small modifications in the size or shape of the beak, which provided a feeding advantage in the competition for food. Natural selection favored those finches with the appropriately modified beak and these small adaptive changes accumulated, generation by generation, ultimately leading to the distinctive variation in beak types observed by Darwin and still seen today.

This sequence of events is typical of an evolutionary phenomenon known as *adaptive radiation;* a species enters a geographic region where there is little competition for resources and a wide variety of available habitats. As population density increases, *intraspecific competition* for food, nesting sites, and other requirements for life becomes intense, causing some members of the species to migrate to different areas where there is less competition for resources. If their initial attempts to colonize the new habitat are successful, the subpopulation will gradually accumulate traits that are better suited to the new set of environmental conditions. Over long periods of time, adaptive changes may accumulate to the point where one or more new species evolve (Figure 3.8).

## The Origin of Species

Darwin's observations during the voyage of the Beagle convinced him that all species of living organisms arose through a slow and gradual accumulation of adaptive characteristics acquired over a long period of time. He concluded that modern species evolved in a slow progression from the extinct forms that preceded them. He could no longer accommodate the

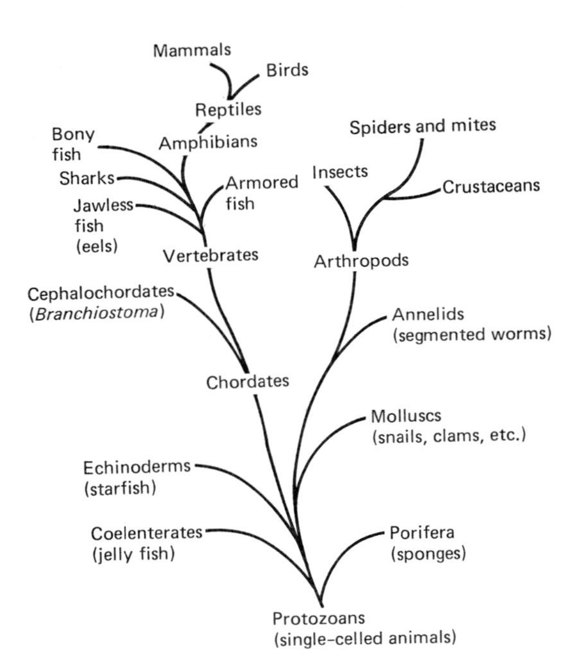


**Figure 3.9** A monophyletic model of animal evolution.

traditional beliefs of special creation and fixity of species with these observations.

Darwin spent over twenty years studying and writing before he finally published *On the Origin of Species* in 1859. This book was a comprehensive and persuasive argument for the theory of organic evolution, and was accepted by the scientific community within a surprisingly short period. Although some prominent theologians mounted major resistance, by the end of the nineteenth century little in the way of a serious scientific challenge to the theory existed.

Darwin's theory of organic evolution has been called "descent with modification." The theory states that each species of plant and animal undergoes continual change, eventually either becoming extinct or giving rise to new and better adapted species. If the theory is extended to its logical conclusion, it predicts that there was one original life-form and that all life-forms that followed derived from that form in a branching pattern of development. This concept of evolution is called the *monophyletic model* (Figure 3.9).

A second pattern of species development is provided by the *polyphyletic model,* which proposes that life originated more than once, with each point of origin serving as the basis for a separate branching evolutionary tree (Figure 3.10). Some religious fundamentalists allow only for variation

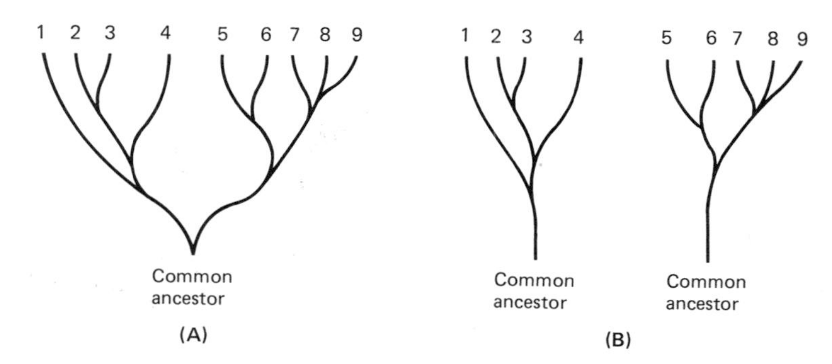


**Figure 3.10** Monophyletic and polyphyletic models of evolution. The monophyletic model (A) pictures life as having had a single point of origin from which all other life forms evolved. The polyphyletic model (B) permits more than one point of origin with each point having given rise to a variety of life forms through separate patterns of divergence.

within lines of descent—for example, the modern horse may be considered a modified descendant of the first created horse, but each is a horse and there has been no evolution from one species to another.

Modern biologists believe that the formation of new species has been a common event in the history of life on Earth and support the monophyletic pattern of species *divergence*. Evolutionary change that results in the formation of new species is called *macroevolution*. Small-scale adaptation within a species population that does not result in species formation is called *microevolution*.

## Problems with the Theory

Darwin recognized certain shortcomings in his theory of organic evolution. One perplexing problem was that the process of evolution requires a tremendously long time; literally millions of years. In Chapter 2 we reviewed the arguments concerning geologic time and found that the time scale generally accepted during the nineteenth century was far short of this requirement.

Another problem with Darwin's theory of organic evolution was that it did not explain exactly *how* and *why* evolution occurred at all. That is, what is the source of biological variation within species, and how are the best adapted variants selected, generation after generation, to produce a new species of plant or animal? The first part of this problem was not solved by Darwin, for it required an elucidation of the mechanics of heredity, knowledge of which was not fully developed until the first decades of the twentieth century (see Chapters 4 and 5).

## Natural Selection

Darwin did see clearly the *process* by which variation within a population of organisms could lead to the accumulation of change. He formalized his ideas in his second major theory, the theory of *natural selection*. Darwin reasoned that, just as had occurred among the finches and tortoises of the Galápagos Islands, nature exerts a formative influence on all species. Natural variation among species provides the raw material by which environmental forces produce adaptive change. The immediate problem was to account for the way in which this natural selective process operates.

Darwin was aware of an analogy that could help him. For many centuries, man has effectively carried out selective plant and animal breeding to improve agricultural varieties and to breed better kinds of dogs for use as ratters, hunters, sheep herders, etc. This practice of *artificial selection* has produced varieties significantly different from the original breeding stock. Thus, it is clearly possible for selective breeding to bring about large-scale change within a species. But the analogy of artificial selection could not be translated directly to natural populations. Artificial selection involves the intelligence of man, who, generation after generation, selects only desirable members of a population as breeding stock. Carrying this analogy too literally to the natural world would strongly imply an intelligence in nature that is goal-oriented, and that evolution proceeds toward this goal. Darwin preferred to seek a more mechanistic explanation.

Darwin spent a great deal of time reading, experimenting, and contemplating ways in which selection might occur in natural populations. In 1838 he happened to read *An Essay on the Principle of Population* by Thomas Malthus. In this essay, Malthus attempted to dispel what he considered to be a false optimism generated by utopian writers of this period. Malthus believed that, despite contemporary advances in agriculture and technology, there was not a period of bountiful rewards awaiting the human species. He marshalled his counterarguments and published them in his essay.

Essentially, the Malthusian argument stated that because of continuing passion between the sexes, there is a tendency for the human population to increase at a rapid rate. This natural tendency toward increase is generally held in check by such populational restraints as fire, war, plague, and famine. Malthus did not foresee improvements in agriculture that would produce enough food for an ever growing human population, the development of modern methods of birth control, or an end to such natural calamities as fire and plague. War, he reasoned, is always with us. All in all, it seemed unlikely to Malthus that the human race was approaching a condition of earthly paradise.

On reading Malthus' essay, Darwin immediately realized that the explanation of natural selection lay in this tract. *All* species of living organisms have far greater reproductive capacities than they ever realize. A single oyster may produce thousands of eggs, but if ten of them should reach maturity, even this small number would greatly exceed the average survival rate. It is only the members of a population that reach reproductive maturity that contribute their hereditary information to the next generation. Thus, those factors that are effective in reducing the potential breeding population of a species will be the very factors that are responsible for natural selection. These selective factors include: amount of water, range of temperature, intensity of light, extent of predation, and availability of nesting space. Within the range of variation displayed by any generation of a species, some variants will be better adapted to survive than others. In the severe competition for resources among the large number of hatching eggs and germinating seeds in nature, only those individuals best adapted to the rigors of the environment will survive to reproduce. The slow and gradual accumulation of adaptive changes over long periods of geologic time results in organic evolution—the creation of new species of organisms.

# 4 Mendelian Genetics

## Introduction

The concept that life begets life is essential to the theory of organic evolution. If life were continually created a priori from inert matter, as was believed in the nineteenth century, organic evolution would be a meaningless concept. In modern scientific thought, life is a continuum, with all living organisms tracing their heritage back to that dim era of Earth history when life originated. The concept of life as a continuum raises two important questions: (1) How is life reproduced? and (2) How is the inherited information of life transmitted through the reproductive process?

## Sexual Reproduction

The reproduction of life occurs through a carefully controlled sequence of events called *sexual reproduction* (Figure 4.1). The basic events of this process are the same in all multicellular organisms. *Gametes* (sperm and egg cells) are produced by the parent organisms and when brought into close proximity unite to form the *zygote* (fertilized egg). The zygote is the first

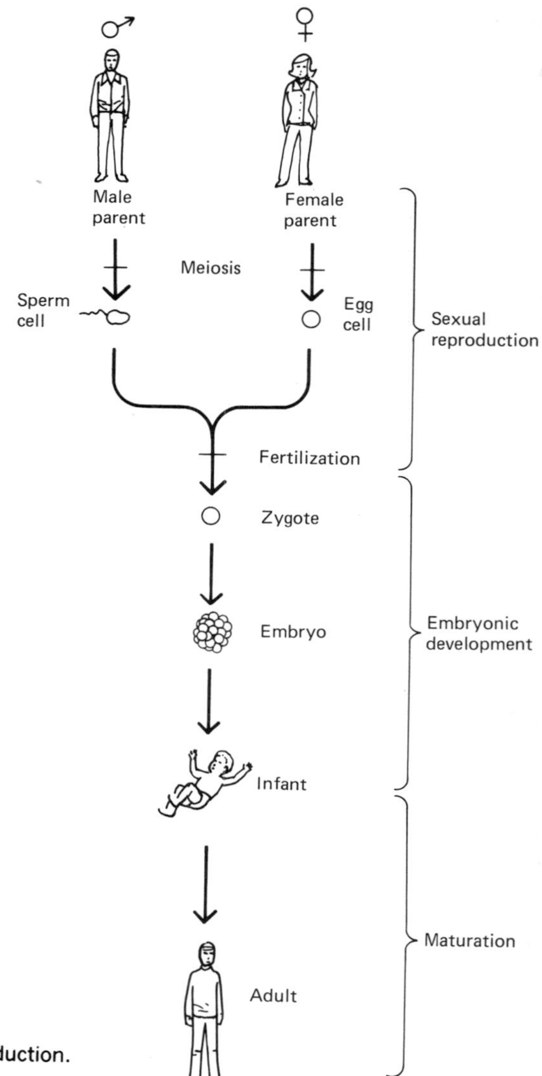


**Figure 4.1** Sexual reproduction.

step in formation of a new individual of the species. The first cellular divisions of the zygote form the *embryo,* which continues to grow by cell division until a fully formed organism is produced. In multicellular organisms the embryonic growth stage is marked by cellular differentiation and specialization leading to the formation of the tissues, organs, and systems typical of the species.

## Information Transfer

Gametes are the only connecting link between generations. The hereditary information passed from parent to offspring is contained in these small bits of living matter. According to modern biology, heredity is controlled by chemical units called *genes* and every heritable trait of an organism is the product of gene action. Genes are contained in the nucleus of the cell, and when sexual reproduction occurs the genes of the parents are transmitted to the zygote through the nuclei of the sperm and egg cells. As a result of the reproductive process, the zygote receives one gene for every trait from each parent. Thus, adult organisms normally contain two genes for each trait.

Two important events occur during sexual reproduction: (1) *gamete segregation,* and, (2) *gene recombination.* Gamete segregation is the separation of the paired genes of the parent cell so that the gamete receives only one gene of each pair. The union of sperm and egg cells restores the paired-gene condition to the zygote. Gamete segregation and fertilization result in gene recombination, i.e., a combination of alleles different from that of either parent.

The mechanism that controls the separation of genes during gamete production is a special type of cell division called *meiosis.* Genes occur in threadlike structures called *chromosomes* and there are two chromosomes of each kind in every adult cell. In the normal sequence of events, each gamete receives one of each pair of chromosomes. Thus, separation of gene pairs in gamete formation (gamete segregation) is accomplished by manipulation of the chromosomes during meiosis.

Each species of organism has a specific number of chromosome pairs. The cell depicted in Figure 4.2 contains two pairs of chromosomes. Each chromosome pair is given a different shape, and one chromosome of each pair is shaded so that individual chromosomes can be followed through the steps of meiosis. Note that with two pairs of chromosomes there are four possible chromosome combinations in the gametes (compare Figures 4.2 and 4.3). Each of these gametic combinations has the same numerical probability of being formed by meiotic cell division, and, when large numbers of gametes are considered, each type will be present in equal frequency.

When fertilization occurs, any sperm cell may unite with any egg cell. Since the union of sperm and egg cells is random, each of the possible zygotic combinations will occur with equal frequency. The fertilized egg grows by means of a second type of cell division called *mitosis.* In mitosis, the chromosome content of every new cell is identical with that of the parent cell, thus guaranteeing that all cells of the adult organism are genetically identical to the zygote. Cells with both chromosomes of each

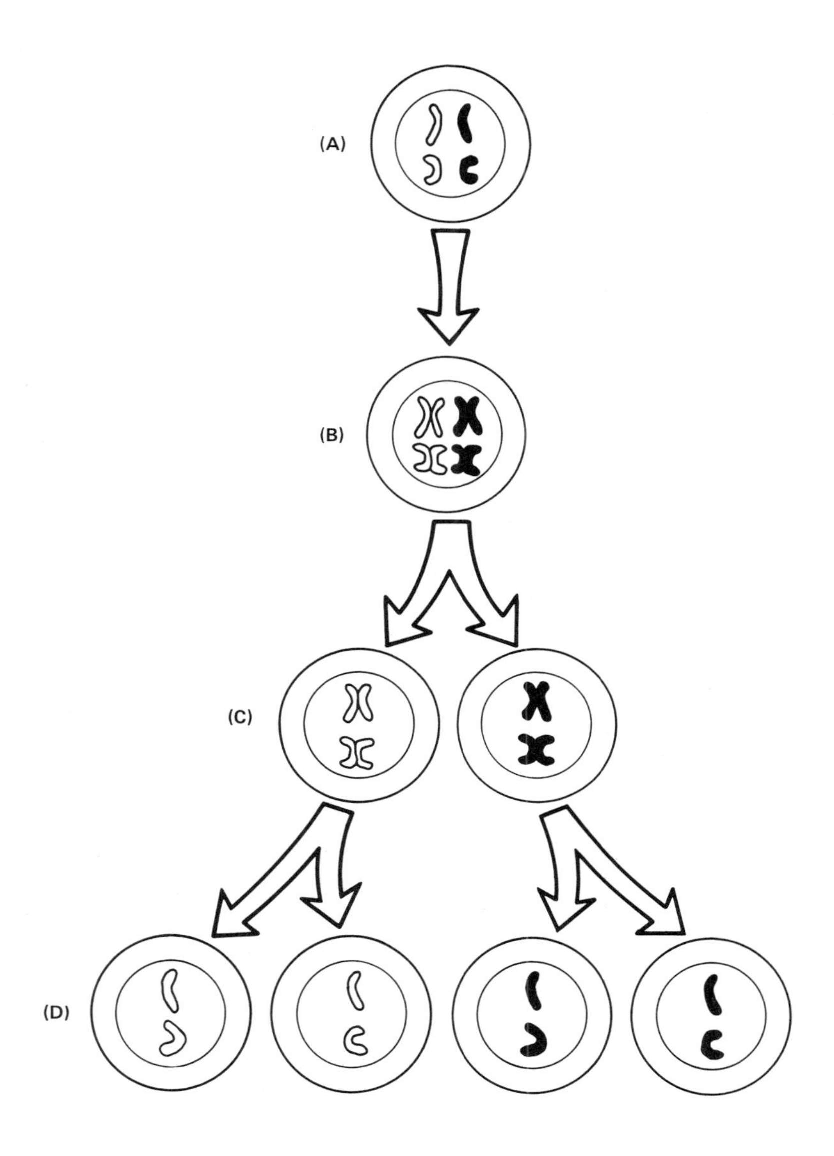


**Figure 4.2** Meiosis. Gametes are formed by meiotic cell division, which is a series of two consecutive cell divisions resulting in the reduction of chromosome number in the gametes. (A) Nondividing cell with two pairs of chromosomes. (B) Chromosomes replicated prior to meiotic division. Replicated chromosomes are considered to be a single chromosome until the two strands separate during the second meiotic division. (C) The first meiotic division separates the chromosome pairs, with one member of each pair going to a different cell. (D) The second meiotic division separates the strands of the replicated chromosomes, producing four gametes, each of which contains *one* chromosome of each pair.

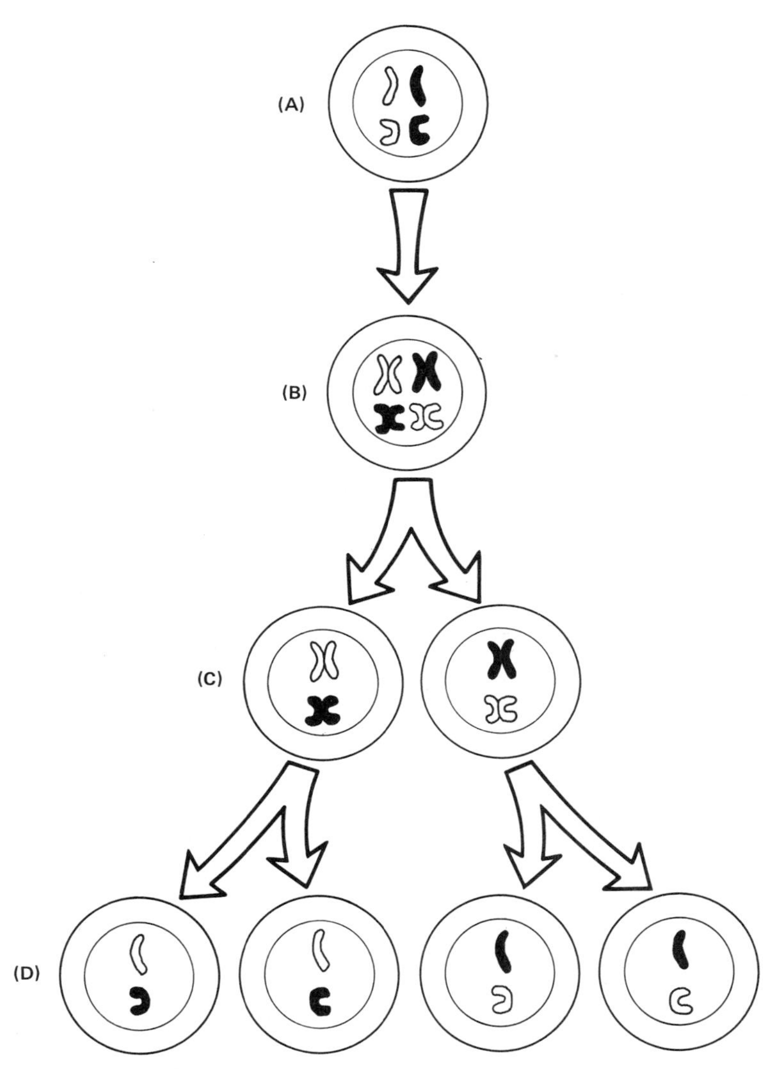


**Figure 4.3** Alternate result of meiosis. Because there is more than one possible arrangement of chromosomes during the first meiotic division, there is more than one possible combination of chromosomes in the gametes. Compare the above sequence with the events pictured in Figure 4.2. (A) Nondividing cell with two pairs of chromosomes. (B) Replication of chromosomes. (C) Results of first cell division. (D) Gametes. (Compare with gametes formed in Figure 4.2.)

pair present are called *diploid cells,* while those with only one chromosome of each pair present are called *haploid cells.* In the human body, for example, all cells are diploid except the gametes, which are haploid.

## The Gene

The gene is composed of a *nucleic acid* called deoxyribonucleic acid (DNA) and each gene contains hundreds of individual nucleotide units (Figure 4.4). Each nucleotide unit is made up of three parts: (1) a ribose sugar unit, (2) a phosphate unit, and, (3) one of four organic bases (adenine, cytosine, guanine, or thymine). In the cell, genes are linked together to form chromosomes, which are composed of the nucleotide units plus proteins.

## Molecular Genetics

Genes influence trait development through their control of the chemistry of the cell. Each gene carries a "code" that the cell "reads" as an order to produce a specific chemical molecule called a *protein*. Proteins are complex molecules made up of smaller units known as *amino acids*. Some twenty amino acids occur in living systems, and the proteins of all organisms are made of various combinations of these same amino acid units. Differences between proteins are the result of variation in the number of amino acids present and the sequence in which they occur. The organic bases of the DNA molecules (adenine, cytosine, guanine, or thymine) carry a coded message that can be compared to the Morse code (Figure 4.5). Morse code makes use of two symbols (dot and dash) in various combinations to spell out meaningful words. The genetic code has four symbols—the four bases of the DNA molecule (we can abbreviate these bases as A, C, G, and T). Three of these bases are needed to code a single amino acid; thus, the DNA base sequence A–A–A will always code the amino acid phenylalanine; C–C–A will always code the amino acid glycine; G–G–T will always code the amino acid proline; and so on. Several hundred amino acids make up a single protein, and for the protein to function normally every amino acid must be in its proper place. Any gain, loss, or change in position of any amino acid may result in malfunction or loss of function of the protein.

Proteins are the direct product of gene action and there are many kinds of proteins in every living cell, some serving as structural "building blocks" of the cell, others functioning in the control of cellular processes. Proteins that control chemical reactions are called *enzymes*, and each enzyme has the ability to mediate a single chemical reaction. All heritable traits of an organism are produced by chemical reactions controlled by one or more enzymes. The differences between molds and men, or cauliflowers and elephants, are genetic differences, as it is the gene that is passed from generation to generation. But, in the development of traits, proteins serve as the agents of the genes and control the myriad chemical

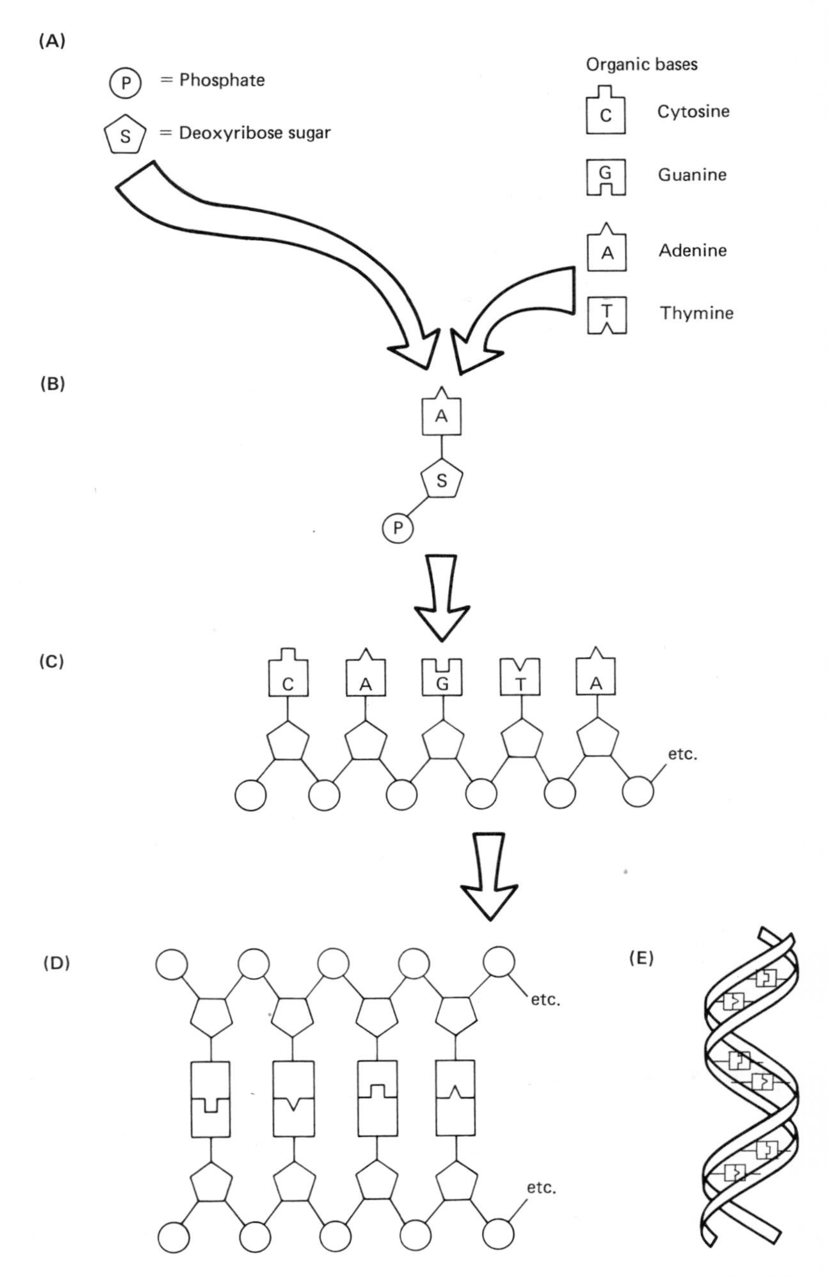


**Figure 4.4** Structure of the gene. (A) Building blocks of DNA nucleotide. (B) A single DNA unit, called a *nucleotide*, is made up of a phosphate molecule, a sugar molecule, and one of the four organic bases. (C) Chain of nucleotide units. A single gene may contain 1000 or more nucleotide units. (D) Chromosomes contain double strands of DNA nucleotides—always linked by cytosine–guanine or adenine–thymine bonds. (E) In the cell, the double strands of DNA molecules take the form of a double helix.

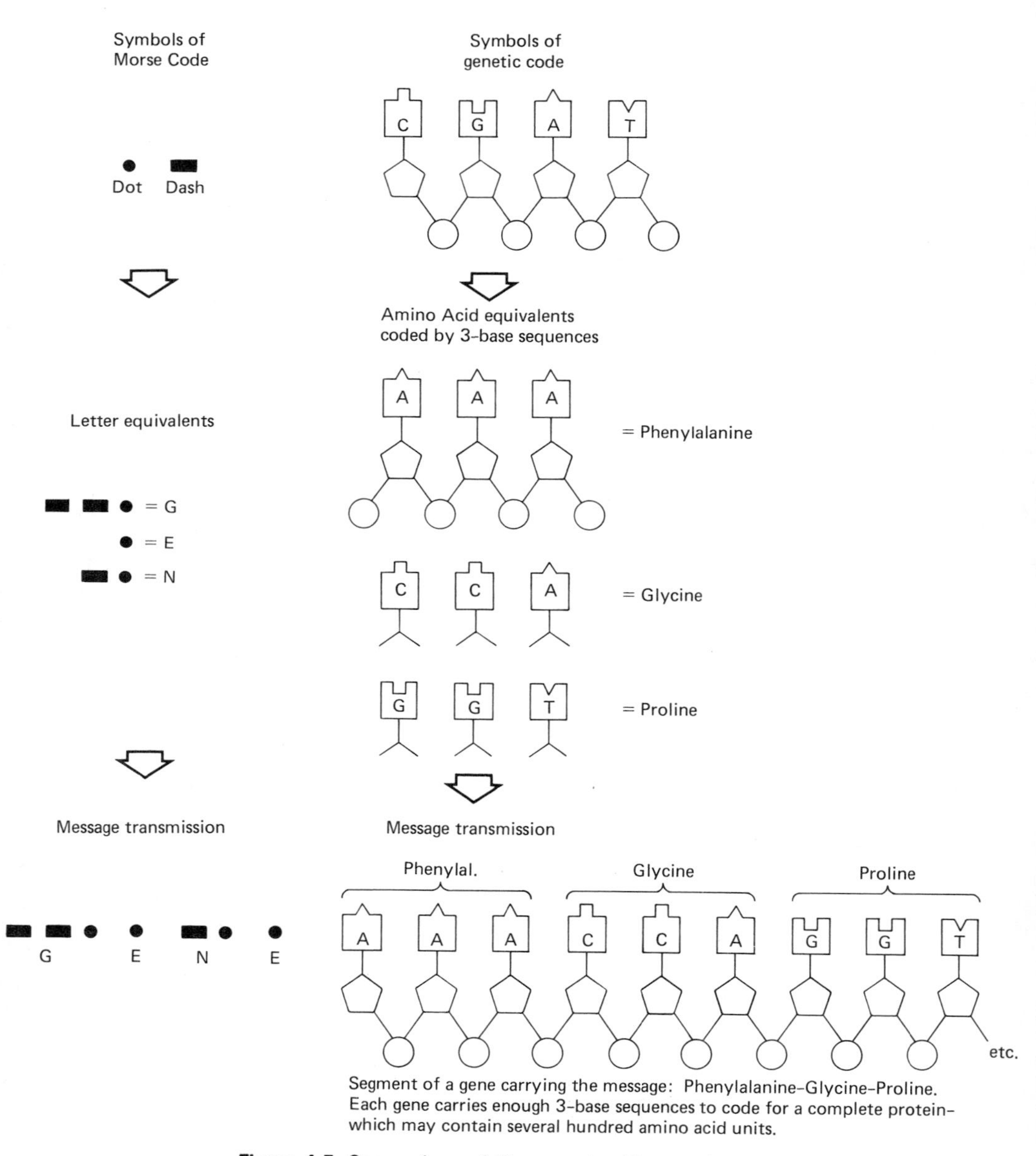


**Figure 4.5** Comparison of Morse code with genetic code.

reactions that determine the final size, shape, and biological capabilities of the new organism.

## Galactosemia

As an example of the way genes function in trait development, we can consider the rare human birth defect called *galactosemia*. Most children can digest galactose sugar because they possess a gene that signals cellular production of an enzyme which digests galactose. However, some children inherit a form of the gene that lacks the ability to produce this enzyme. When these children drink milk, galactose present in the milk accumulates in their blood streams, saturates body tissues, and causes developmental abnormalities. Unless these children are placed on a milk-free diet very early in life, liver damage, cataracts, and mental retardation can result.

Two forms of the gene for production of the galactose-digesting enzyme exist. We can designate these genes as *G* and *g*. The gene *G* functions normally and cells that contain this gene produce sufficient enzyme for galactose digestion. Recalling that every individual has two genes for every trait, normal children may have either of two gene combinations: *GG* or *Gg*. The *g* form of the gene lacks the ability to produce the galactose-digesting enzyme and individuals who have two *g* genes (*gg*) cannot digest galactose and will develop galactosemia unless proper dietetic restrictions are imposed.

## Definitions and Concepts of Genetics

The two forms of a gene pair that control a specific trait are called *alleles* (singular, *allele*). Each pair of alleles occupies a specific site on a chromosome called the gene *locus*. Many genes occur in two allelic forms, as the *G* and *g* forms of the gene for production of the galactose-digesting enzyme. When both allelic forms of the gene are present in a single cell, one gene of the pair often dominates the other. For example, in galactosemia, individuals with the gene pair *Gg* are normal because the *G* allele produces sufficient enzyme for galactose digestion. In this situation the *G* allele is *dominant* to the *g* allele. Because the *g* allele is not expressed, it is said to be *recessive* to the dominant allele.

Another example of the dominant–recessive gene relationship is albinism in humans. The development of the pigment melanin is controlled by a single pair of genes (*A* and *a*). Individuals who have either gene pair *AA* or *Aa* have normal skin, hair, and eye pigmentation because the *A* allele causes production of the enzyme necessary for melanin synthesis. The *a* allele lacks the ability to produce this enzyme, and individuals who are genetically *aa* have no pigment formation and will be albino. In this example, the *A* allele is dominant and the *a* allele is recessive.

When both alleles are the same (*AA* or *aa*), an individual is pure line or *homozygous.* Individuals with alternate forms of the gene (*Aa*) are *heterozygous.* An individual with two dominant alleles is homozygous dominant; and an individual with two recessive alleles is homozygous recessive.

## Breeding Crosses

"Breeding true," to the plant or animal breeder, means that a cross between parents having the same trait produces only offspring that resemble the parents. It has long been recognized that some traits "breed true" while others do not. The reason for this inconsistency was not known until the 1880s when the Austrian monk, Gregor Mendel, set up carefully controlled breeding experiments using garden peas. From his research, Mendel derived the fundamental principles of heredity and provided an explanation for these unexpected results. Because of his pioneering work the basic principles of heredity are often called "Mendelian genetics," although the concept of the gene was not introduced into biology until long after Mendel's death. Today, the use of the breeding testcross is an integral part of genetic research.

Crosses between two black sheep will always produce 100% black lambs; i.e., black coat color breeds true. In crosses between white sheep the situation is different. One testcross between white-coated parents may produce 100% white offspring whereas another testcross between different white-coated parents may result in a mixture of black and white lambs. Today, using the principles of Mendelian genetics, it is not only possible to explain these results but to predict the probability of black or white lambs being produced by particular parents.

A cross between homozygous white sheep (*WW* x *WW*) will always produce 100% white lambs (Figure 4.6A). Because all black sheep are homozygous (*ww*), crosses between black parents will always produce 100% black lambs (Figure 4.6B). Thus, as long as crosses are made between homozygous individuals who have the same form of trait expression, the trait will breed true. But what is the result of a mating between a homozygous white sheep and a black sheep?

The offspring of a cross between a pure-line white sheep (*WW*) and a pure-line black sheep (*ww*) will be 100% white-coated (Figure 4.7A), but the individuals of the first generation (called the $f_1$) are not genetically like either parent. All the $f_1$ offspring are heterozygous (*Ww*) white sheep be-

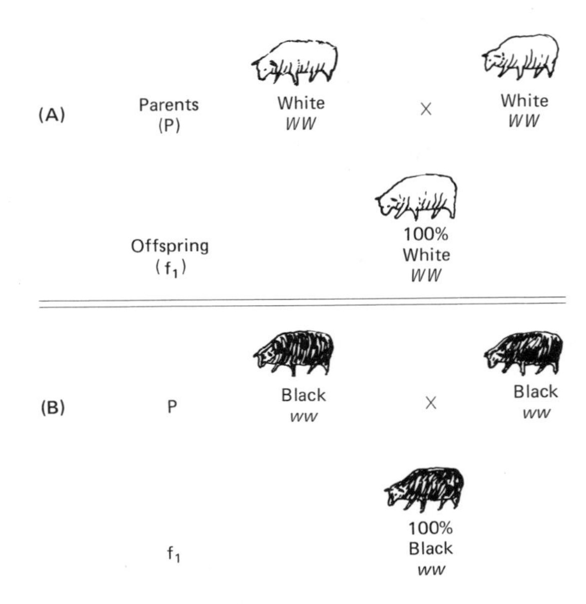


**Figure 4.6** Breeding crosses. (A) Between two homozygous white sheep, and (B) between two homozygous black sheep.

cause the allele for white coat color is dominant to the allele for black coat color. Note that genetic diversity is greater in the offspring than in the parents (i.e., the offspring are heterozygous (*Ww*) while the parents are homozygous). Further crosses can be made between white-coated heterozygous sheep of the $f_1$ generation (Figure 4.7B). Through gamete segregation, one-half of all the sperm cells produced by the ram contain the dominant allele (*W*) and one-half contain the recessive allele (*w*). The same genetic ratio applies to egg cells produced by the ewe. The recombination of these gametes is entirely random—any sperm and any egg cell have an equal opportunity to unite in fertilization.

The simplest way to determine the genetic combinations that will result from the mating of $f_1$ individuals is to set up a table called the *Punnett square* (Figure 4.7B). In the Punnett square, the genetic types of sperm cells are entered at the top of the table and the genetic types of egg cells along the left side. The square is completed by entering the genetic contributions of the sperm and egg cells in the blanks of the square (each completed square represents a zygote). When the squares have been completed we find that there are four possible combinations of sperm and egg cells: *WW*, *Ww*, *wW*, and *ww*. Note that *two combinations* will produce the heterozygous condition (*Ww*, *wW*); therefore, there will be twice as many heterozygous lambs as either homozygote.

In the second generation ($f_2$) one-fourth of all individuals will be ho-

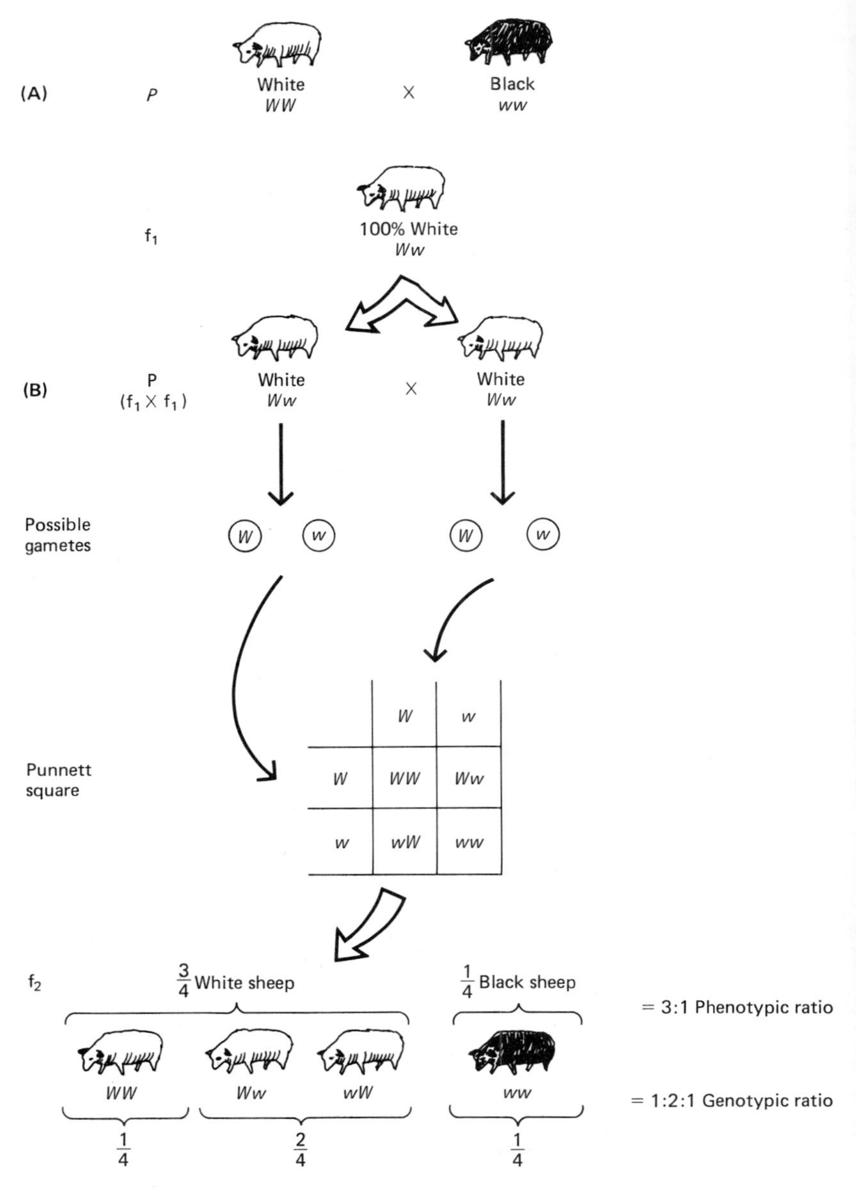


**Figure 4.7** Breeding crosses. (A) Between homozygous white and homozygous black sheep, and (B) between heterozygous white sheep.

mozygous dominant (*WW*); two-fourths (or one-half) of all individuals will be heterozygous (*Ww*); and one-fourth of all individuals will be homozygous recessive (*ww*). Considering the frequency of *trait* expression, three-fourths of the $f_2$ generation will have at least one dominant allele and will have white coat color (sum of *WW* plus *Ww*), and one-fourth will be homozygous recessive with black coat color.

The term *genotype* refers to the genes actually present in the zygote. Thus, white sheep may have either of two genotypes: *WW* or *Ww*. All black sheep will have the genotype *ww*. The term *phenotype* refers to the trait actually seen. In this example only two phenotypes are possible; white coat color and black coat color.

The mathematical probability that a certain event will occur can be expressed as a ratio. In genetics, the probability that a particular mating will produce offspring of a certain genotype is called the *genotypic ratio.* Thus, the genotypic ratio of the $f_2$ generation in Figure 4.7B is 1 : 2 : 1 (1/4 *WW* : 2/4 *Ww* : 1/4 *ww*). The probability of certain phenotypes occurring is expressed as the *phenotypic ratio.* The phenotypic ratio in Figure 4.7B is 3:1 (3/4 white-coated and 1/4 black-coated). These ratios are statistical statements and require large numbers of offspring to be valid.

## Lack of Dominance

Not all traits follow the dominant–recessive inheritance pattern. Some hybrids display a degree of trait expression that is intermediate to the traits of the parents. This type of genetic relationship is called *lack of dominance.* An example of this condition is flower-color inheritance in snapdragons.

If a pure-line, red-flowered snapdragon (*RR*) is crossed with a pure-line, white-flowered snapdragon (*rr*), all offspring will be heterozygous pink-flowered, (Figure 4.8). Since there is no dominant or recessive gene for flower color in snapdragons the heterozygous condition results in a form of trait expression intermediate to the pure-line parents.

When there is lack of dominance, the cross of hybrids produces a phenotypic ratio that is identical with the genotypic ratio. A cross between two pink-flowered snapdragons (*Rr*) produces an $f_2$ genotypic ratio of 1 : 2 : 1 but, with the distant phenotypic expression of the heterozygotes, the $f_2$ phenotypic ratio is also 1 : 2 : 1.

## Multiple Allelism

The gene produces variation through its control of the chemical function of the cells of the organism. If gene alleles occur in a dominant–recessive gene relationship, two phenotypes are possible. When there is lack of dominance, three phenotypes are possible. Other genetic relationships, such as multiple allelism, produce an even wider range of phenotypic variability.

In multiple allelism there are more than two possible forms of the gene. Allelic forms of a gene are produced through gene *mutation*. The result of gene mutation is that a new form of the gene is created. The original gene is called the "wild type" and the new gene the mutant. If either

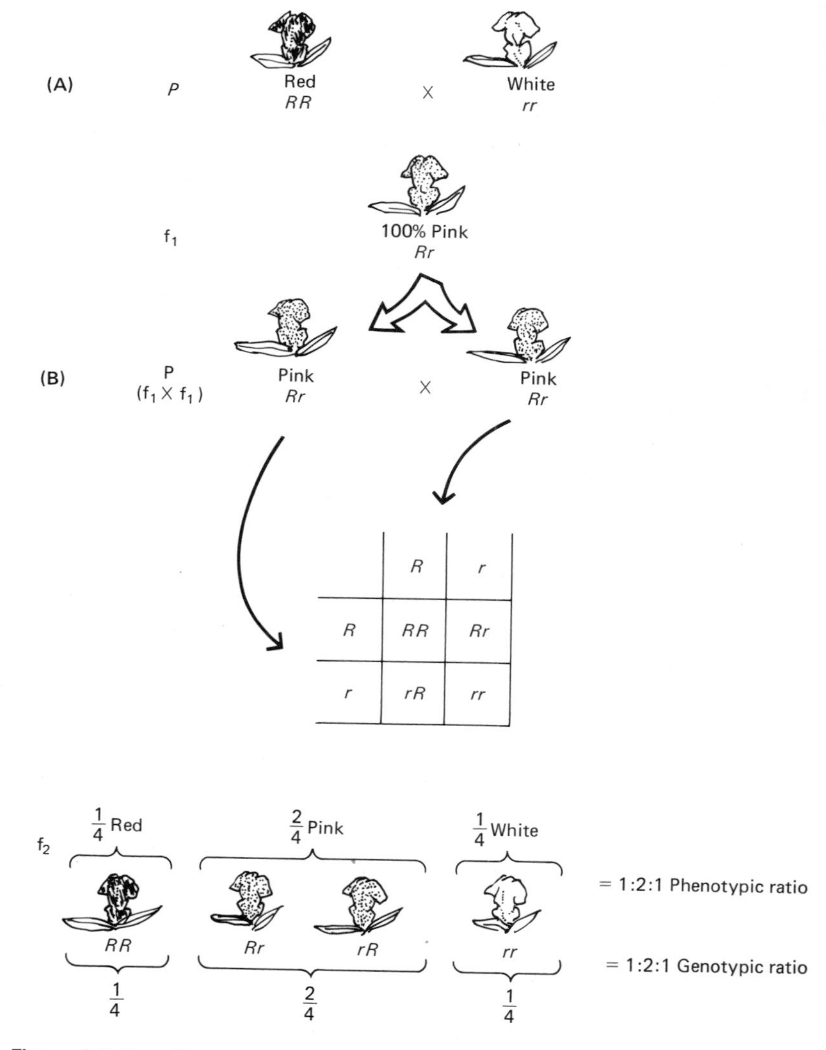


**Figure 4.8** Breeding cross with lack of dominance.

(A) Alleles for coat-color inheritance in rabbits

$C$-Full coat color

$c^{ch}$ -Chinchilla coat pattern

$c^{h}$-Himalayan coat pattern

$c$-Albino

(B) Order of dominance among coat-color alleles
(read arrow ⟶ "is dominant to")

$C \longrightarrow c^{ch} \longrightarrow c^{h} \longrightarrow c$

(C) Possible genotypes and phenotypes

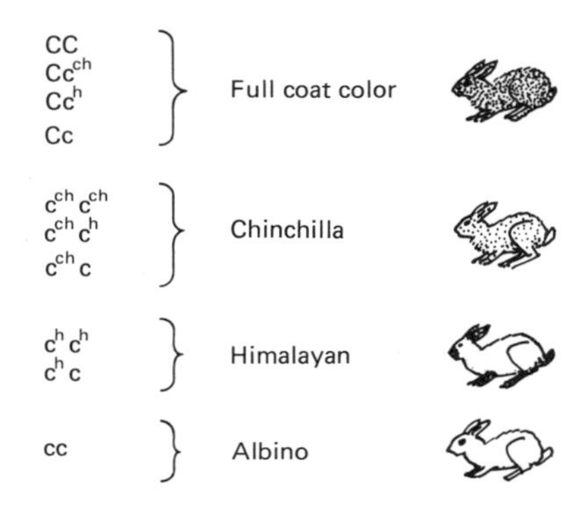


**Figure 4.9** Multiple allelism and coat-color inheritance in rabbits.

of these two genes should undergo mutation, a third allelic form of the gene is produced, and as many as 200 allelic forms are known for some genes. Although a large number of possible alleles may exist, it should be kept in mind that *only two genes* (or alleles) for a trait can normally be present in the cells of any organism.

There are four allelic forms of the gene that controls the distribution of coat color in rabbits. Figure 4.9 presents the genetic relationship among these genes, the possible genotypes, and the resulting phenotypic expressions. The significant thing to note is that there are ten possible genotypes and four possible phenotypes. Multiple allelism increases both genotypic and phenotypic variability.

# Mutation

The DNA that makes up genes is a very stable chemical; however, spontaneous changes do occur in DNA and the result is gene mutation. A mutation is an inheritable change in a gene. Viewed from the perspective of the coding function of DNA molecules, a mutation is a change in the base sequence of a gene resulting in alteration of the amino acid content of the protein molecule it codes. A deviation in a single DNA base will cause a mutation.

Changes in the amino acid sequence of proteins generally have a deleterious effect on the protein molecule. Often, the result of gene mutation is the production of a nonfunctional enzyme, as seen in our discussion of galactosemia, where the mutant gene produces a nonfunctional enzyme molecule. Another possibility is that an altered enzyme will have a chemical function different from the enzyme produced by the normal gene. *Sickle-cell anemia,* a genetic disease most common in American and African Blacks, is caused by a mutant gene that produces an aberrant form of the hemoglobin molecule. Since the hemoglobin molecule is essential for transport of oxygen from the lungs to the tissues of the body, individuals with abnormal hemoglobin molecules develop a severe form of anemia. In this instance, the gene has not lost the ability to function but, because of the altered base sequence, produces an enzyme whose function is harmful to the organism.

Genetically, sickle-cell anemia provides another example of lack of dominance; the *SS* individual produces 100% normal hemoglobin; the *ss* individual produces 100% abnormal hemoglobin; and the red blood cells of the *Ss* individual contain one-half normal and one-half abnormal hemoglobin. Individuals with the *ss* genotype suffer from bouts of severe anemia brought on by the lowered ability of the abnormal (sickled) hemoglobin to transport oxygen—especially when the afflicted individual is physically active. The heterozygous individual (*Ss*), having one-half abnormal hemoglobin in the red blood cell, suffers from a milder form of anemia. Curiously, the heterozygote has an adaptive advantage over the other two genotypes in malarial regions because the protozoans that cause malaria cannot live in red blood cells that contain abnormal hemoglobin. Thus, the *ss* individual suffers from severe anemia, the *SS* individual is susceptible to sickness and death from malaria, and the *Ss* individual has only mild symptoms of anemia. In nonmalarial regions the selective advantage of the *Ss* genotype is lost and the *SS* genotype is the healthier. The survival value of a mutant gene is not always either positive or negative but may be dependent upon the environment in which the organism lives.

Most gene mutations are harmful. The genes present in any living organism are the result of a long period of selective adaptation to environmental pressures and the probability that a random mutation will prove beneficial to the organism is remote. Occasionally, beneficial mutations do occur and these mutations are very important to the progress of organic evolution. Specific examples of beneficial mutations are presented in Chapter 7.

As most mutations are recessive, they will not be expressed in any member of a population until they occur in the homozygous condition. For this reason, recessive mutations may remain in a population for many generations before being expressed. Once phenotypic expression does occur, natural selection determines the fate of the gene. Lethal mutations, which result in the death of the organism, provide a clear example of the fate of mutants in a population. Lethals are continually eliminated from the population, but even this extreme selection does not result in complete elimination of the lethal gene. As long as new mutations occur the gene will be present, and heterozygous individuals, often called gene *carriers*, will maintain detrimental genes in the population. Thus, mutant genes of negative selection value are a constant part of the genetic makeup of any population.

Beneficial genes, on the other hand, are favored by natural selection. Beneficial recessive mutations move through a population rather slowly, but if the mutation has survival value the frequency of the mutant gene will increase at the expense of the original genetic form.

Dominant mutations, although rare, do occur and these traits appear immediately. The human birth defect achondroplasia (a severe form of dwarfism) is an example of a trait caused by a dominant mutation. Another example of a dominant mutation is the occurrence of a dark (melanic) form of insect (discussed in Chapter 7).

## Mutation Rate

Gene mutations occur spontaneously without apparent cause; nevertheless, they do occur at a predictable rate. The mutation rates for some gene mutations are presented in Table 4.1.

**TABLE 4.1 Mutation Rates**

| ORGANISM | TRAIT | MUTATION RATE |
|---|---|---|
| Bacterium (*Escherichia coli*) | Streptomycin resistance | 1 in 1,000,000,000 |
| Fruit Fly (*Drosophila melanogaster*) | White eye | 4 in 100,000 |
| Humans | Achondroplasia | 2.8 in 100,000 |
| | Albinism | 2.8 in 100,000 |
| | Hemophilia | 3.2 in 100,000 |
| | Total color blindness | 2.0 in 100,000 |

Certain factors can cause an increase in mutation rate above the spontaneous level. High-energy radiation, such as cosmic rays, X-rays or radioactive-decay products (e.g., gamma rays) are thought to increase the rate of mutation in a linear fashion. That is, any increase in the intensity of radiation will cause a concomitant increase in mutation rate. Since most mutations are potentially harmful to the organisms in which they occur, exposure to even small amounts of high-energy radiation should be avoided. Many chemicals, such as some pesticides, nitrites (food preservatives), and drugs are also thought to cause gene mutations.

## Role of Mutation in Organic Evolution

Mutation has been called the raw material of evolution. The greatest source of genetic diversity is the recombination of existing genes, but gene recombination alone will produce only variation within existing traits. Mutation, on the other hand, provides the possibility that the mutant gene will produce a new trait. This event is extremely rare, but the course of evolution has been charted by such uncommon and random events occurring over an enormous amount of time.

# 5 Population Genetics

## Introduction

Mendelian genetics is concerned with the hereditary factors of individuals, but the modern theory of organic evolution stresses the fate of genes in populations. An individual will live or die; a species will either adapt to environmental change and survive, or it will become extinct. In the course of adaptive change, new species of organisms may be formed. The survival of the species, and the formation of new species, is determined by the genetic response of the population as a whole, and not by the life or death of any one organism.

## Populations and Gene Pools

Because the population is important in evolution, we must understand the populational principles derived from Mendelian genetics. A *population* is a group of organisms of a single species. It may include all members of a species alive at any one time or refer to all members of the species occupying a specific geographic area. For example, we can consider the human population of Los Angeles, the United States, or the world.

All the genes present in a population represent the *gene pool* of that population. Events that occur within the gene pool determine the response of a population to the environment in which it lives. The major sources of variability within the gene pool are gene recombination and mutation—which are exposed to the environment as phenotypic variation. Selection favors adaptive phenotypes and tends to eliminate nonadaptive phenotypes. The result of this interaction is a change in the relative numbers of genes conferring adaptive and nonadaptive traits.

There are abundant data correlating coat color in mammals with predominant background colors found in their habitat. Mammals living in arctic regions generally have white coats, deer of the woodlands have tawny coats, and the bold stripes of the tiger blend well with patterns of shade and sunlight found in the tropical jungle. In such cases the color-correspondence of the animal with its habitat provides a form of concealment—giving protection to the prey and a hunting advantage to the predator. Such adaptive coloration is the result of natural selection operating on variations in coat color over a long period of time, eventually producing a population of animals that have the advantage of being camouflaged against detection when in their natural habitat. A classic example of the role of natural selection in determination of coat color is provided by the studies of Lee R. Dice of the University of Michigan. Dice studied populations of mice occupying adjacent habitats of dark-colored basaltic rocks and light-colored sand. The mouse population of the light-sand area was uniformly light-colored, matching the sand, while the mice found in the dark-colored rock habitat had a dark coat color. This variation in coat color is explained on the basis of selective predation. Mouse populations have the genetic potential to produce offspring of various coat colors, but on the light-colored sand any offspring with dark coat color would contrast sharply with the background and be easily detected by such predators as hawks, owls, or coyotes. The opposite situation would maintain on the dark basaltic rocks—individuals with light coat color being detected and killed more readily than those with dark coat color. The significance of such selection goes beyond the life or death of the individual animal, because by killing a light-coated mouse the predator has prevented that individual from reproducing and passing its genetic potential to future generations.

Let us assume that coat color in rodents is controlled by a single pair of alleles, *S* and *s*, and that the dominant allele (*S*) confers dark coat color, while the recessive allele (*s*) produces a light coat color when present in the homozygous condition. In the dark habitat the *SS* and *Ss* genotypes would predominate, while in the light-sand environment the *ss* genotypes would make up most of the population.

## Gene Frequency

The *frequency* of a gene is its numerical incidence in the gene pool. In the dark-rock habitat described above, the frequency of the *S* allele increases while the frequency of the *s* allele decreases. In humans, the dominant allele for pigment production has a frequency of 99% and the recessive allele a frequency of 1%. This means that 99 out of every 100 alleles for pigment production is a dominant allele and only one in 100 is recessive. These values are generally expressed as the decimal equivalent of the percentage value; the frequency of the dominant allele is 0.99 and the recessive 0.01.

Because the gene symbols *A* and *a*, or *S* and *s*, are commonly used in Mendelian genetics to represent individual alleles, population genetics uses symbols which stand for *all* of a particular allele present in the gene pool. The symbol $p$ is conventionally used to represent *all* dominant alleles for a particular trait and the symbol $q$ to stand for *all* recessive alleles for the same trait. In the case of albinism, $p = 0.99$ and $q = 0.01$. The total of $p$ and $q$ for any pair of alleles must add up to 1.0 (100%).

The equation

$$p + q = 1.0$$

is a generalized statement that stands for the frequency of any two gene alleles in a gene pool. In the case of albinism

$$\begin{aligned} p &= 0.99 \\ q &= 0.01 \\ \hline p + q &= 1.00 \end{aligned}$$

If the numerical value of either gene allele is known, the frequency of the other allele can be calculated by simple arithmetic:

$$\text{If } q = 0.4 \text{ and } p + q = 1.0$$

then,

$$\begin{aligned} p &= 1.0 - q \\ p &= 1.0 - 0.4 = 0.6 \end{aligned}$$

## Genotype Frequency

The genotype of the organism determines the phenotype, and the symbols $p$ and $q$ by themselves convey no information concerning the number of individuals of different genotypes occurring in a population. However, if the frequencies of individual alleles are known, it is possible to determine the frequency of genotypes. The equation used to determine genotype frequencies in a population is the fundamental tool of population genetics.

In 1908 H. H. Hardy and W. Weinberg independently discovered that the frequency of genotypes in a population can be treated as a special case of the binomial expansion of algebra. Using the symbols $p$ and $q$, this equation can be written:

$$p^2 + 2pq + q^2 = 1.0$$

In this equation, $p^2$ stands for *all* individuals with the homozygous dominant genotype; $q^2$ stands for *all* individuals with the homozygous recessive genotype; and $2pq$ stands for *all* heterozygous individuals (Table 5.1). If these three groups are added together, the total must be 100% or its decimal equivalent 1.0.

**TABLE 5.1 Individual and Populational Symbols**

| INDIVIDUAL | POPULATIONAL | |
|---|---|---|
| *AA* | $pp$ or $p^2$ | (homozygous dominant) |
| *Aa*<br>*aA* | $2\ pq$ | (heterozygous) |
| *aa* | $qq$ or $q^2$ | (homozygous recessive) |

Returning to the example of albinism, we can illustrate the way in which the Hardy–Weinberg equation can be used to determine population genotype frequencies in Table 5.2.

**TABLE 5.2 Albinism**

| GENE FREQUENCIES FOR ALBINISM | GENOTYPE FREQUENCIES FOR ALBINISM | |
|---|---|---|
| $p = 0.99$ | $p^2 = (0.99)^2$ | $= 0.9801$ |
| $q = 0.01$ | $2pq = 2(0.99)\ (0.01)$ | $= 0.0198$ |
| | $q^2 = (0.01)^2$ | $= 0.0001$ |
| | $p^2 + 2pq + q^2$ | $= 1.0000$ |

It can be calculated that 98 out of every 100 individuals in a human population are homozygous dominant for pigment production and that only one person in 10,000 is albino. The frequency of the heterozygote is 2 per hundred; that is, about one person in fifty carries the gene for the albino trait. An important observation derived from population genetics is that the heterozygous condition is much more prevalent than the actual expression of the recessive phenotype would indicate. Figure 5.1 shows the relationship of heterozygosity to the frequency of the two gene alleles. The high level of heterozygosity present in most gene pools thus provides a vast reservoir of hidden variability.

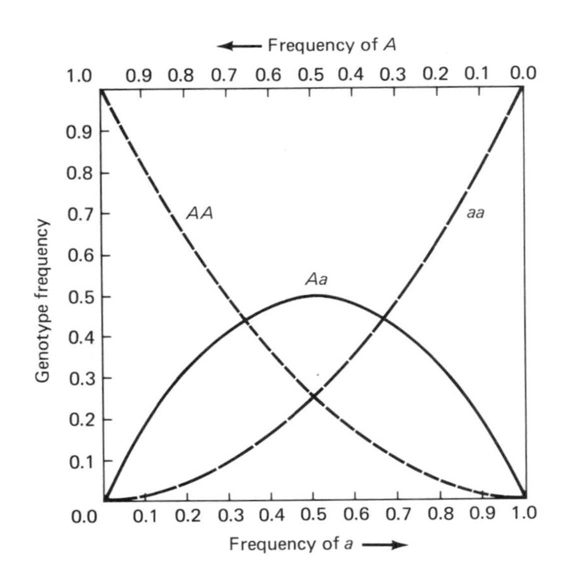


**Figure 5.1** Relationship between gene frequencies and genotype frequencies.

# The Hardy–Weinberg Equilibrium

The equation for genotype frequency developed by Hardy and Weinberg is an *idealized* model of a genetic population because it represents a gene pool in which no changes in gene frequencies are occurring. Such a gene pool is in equilibrium. The important function of the Hardy–Weinberg equation is that it provides an understanding of the equilibrium situation. When actual changes in gene frequencies are observed, both the cause of change and the rate of change can be studied.

# Factors Affecting Gene Frequencies

Gene pool equilibrium can exist only under very strict conditions, and such factors as mutation, natural selection, nonrandom breeding, gene flow, and genetic drift can cause changes in gene frequencies. Gene mutation is the change of a gene from one form to another, and so the very act of mutation results in a change in gene frequency, and, as in the example of coat color in rodents, natural selection tends to increase the frequency of adaptive alleles and decrease the frequency of nonadaptive alleles.

Random mating—the situation in which any reproductive member of a population has the opportunity to mate with any reproductive member of the opposite sex—is also required for gene pool stability. Any factor that restricts random breeding will cause shifts in gene frequencies. In human populations, social, educational, economic, and geographic factors tend to cause nonrandom breeding with the result that differences in gene frequencies occur in subpopulations of the human species. The behavior of some animals in which a few dominant males have exclusive herds of females excludes some individuals from the breeding population and causes changes in the gene pool from generation to generation.

Because of variations in environmental conditions, geographically isolated gene pools of a single species will show differences in gene frequencies. This effect can be observed in the distribution of the gene causing the sickle form of the hemoglobin molecule (see Chapter 4). This gene appears to have originated in Africa and has its highest frequency in black populations of that continent. The highest frequency of the "sickling gene" appears in Central Africa (10–20 percent) and is somewhat lower in North Africa. The gene is nearly absent in most other geographic areas.

The appearance of the sickling gene in populations outside Africa is explained by the migrations of black African carriers to other geographic regions. Change in gene frequencies resulting from the movement of individuals between geographically distinct populations is called *gene flow.*

A final factor influencing the numerical incidence of genes is *genetic drift.* In large populations, chance events play a relatively minor role in determination of gene frequencies, but in small populations chance events, such as minor fluctuations in mate selection, accident, or natural catastrophe, may be very important. For example, in a population of 1000 a specific mutation ($A$ to $a$) occurs in one individual per generation. Chance events may prevent this individual from reaching reproductive maturity, from successfully mating, or from producing viable, fertile offspring, and thus the mutant gene is lost from the population *regardless of its adaptive value.* However, with the same mutation rate a population of 1,000,000 would produce 1000 individuals per generation who possess the mutant gene, and it is much less likely that chance alone would control the fate of this gene. Instead, in the larger population, natural selection would be the most important factor in survival of the mutant form.

Thus, for a gene pool to be in equilibrium the population must be large, there must be random mating, and there can be no mutation, natural selection, or gene flow. When gene frequencies are observed to change, one or more of these factors is operating to create that change.

## Mutation and Genetic Load

Most mutations are recessive and most mutant alleles are deleterious when expressed in the homozygous recessive condition. The negative ef-

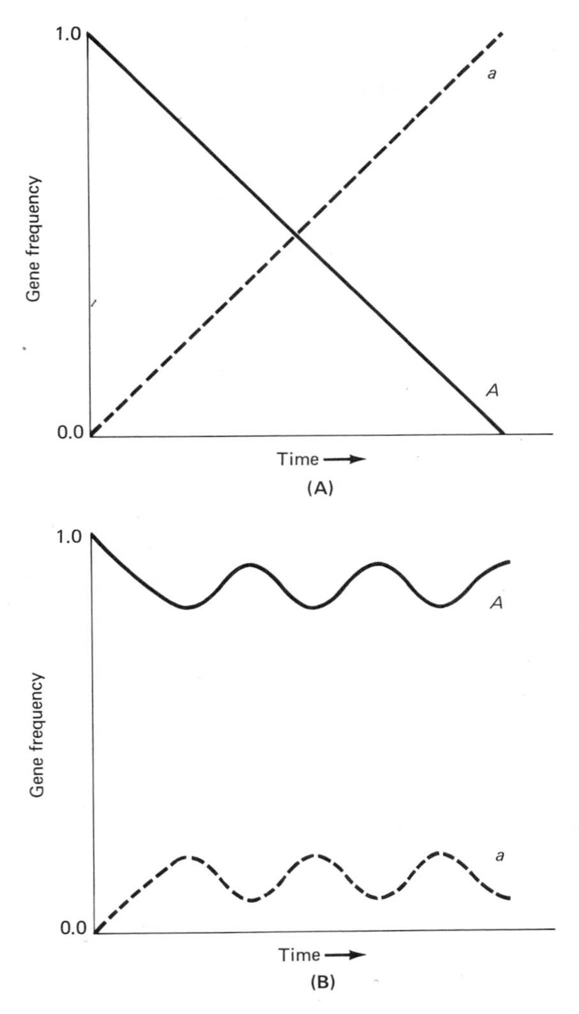


**Figure 5.2** Effect of mutation and natural selection on gene frequencies. (A) Mutation of *A*-to-*a* without selection. (B) Mutation of *A*-to-*a* with selection (*aa* lethal).

fects of such mutations range from slight metabolic disorders to conditions that are lethal. Lethal mutations may result in death during embryonic development or may not take effect until some time after birth. The importance of deleterious genes to the process of organic evolution is in the extent to which they reduce the likelihood that an individual will be successful in reproduction. Any factor that reduces fertility or prevents reproduction will have an effect on gene frequencies.

The effect of a new mutation in a gene pool that is otherwise in equilibrium is presented in Figure 5.2A. If we assume that neither the dominant nor the recessive allele is favored by natural selection, mutation alone causes an even rate of increase in the frequency of the recessive gene and a corresponding decrease in the frequency of the dominant allele. Changes in genotype frequencies that result from such changes in gene frequency appear in Figure 5.1. Note especially the pattern followed by the heterozygous genotype. The maximum frequency of the heterozygote occurs when the frequencies of the *A* and *a* alleles are at 0.5 (50% of the gene pool).

Now let us consider the situation in which the homozygous recessive genotype produces a lethal condition—death occurs before the individual reaches reproductive age. This situation is graphed in Figure 5.2B. The frequency of the recessive allele does not continue to increase as it did in Figure 5.2A; but neither does it disappear from the population, even though its effect is lethal. Instead, the gene, and the genotype it produces, establishes an equilibrium value that is a balance between the rate of mutation and the rate of elimination of the gene through natural selection. Two factors prevent the frequency of the recessive gene from falling to zero: (1) continued mutation, and (2) the maintenance of the recessive allele in the heterozygous condition.

Heterozygous individuals are called gene carriers. The use of the term in genetics is analogous to its use in medicine, where it refers to individuals who carry germs for a disease without themselves developing symptoms of the disease. Because of the carrier effect and continuing mutation, all gene pools contain a number of recessive alleles that, if present in the homozygous recessive condition, will reduce the vigor or cause the death of the individual. This residue of deleterious genes is called *genetic load.* Any factor that increases the rate of mutation will increase the genetic load of the population.

## Species Formation

An important event in organic evolution is the formation of a new species. The definition of a species is traditionally based on similarity and difference; in this sense a species is a population of organisms with obviously similar traits that are transmitted from generation to generation. Two similar species are separated by the fact that the range of variation between the two species is greater than the range of variation within each individual species. Such a definition is obviously fraught with interpretive perils, but to the scientist it *is* meaningful. For the paleontologist, working primarily with the fossil remains of extinct organisms, it is the only kind of definition possible.

A species can also be defined as a group of actually or potentially interbreeding organisms that can produce viable, fertile offspring. The emphasis in this definition is not only on reproduction but on continuity as well. The horse and the ass can mate successfully and produce a viable offspring, the mule; however, the mule itself is infertile and represents a reproductive and evolutionary dead end. For this reason, biologists assign the horse and the ass to two different species.

The reproductive criterion for species definition helps explain the process by which species are formed. The first step in speciation is *geographic isolation* of the species population into physically separated gene pools. In time, events within each gene pool produce *isolating mechanisms* that prevent interbreeding between members of the subpopulations. When these gene pools become *reproductively isolated,* new species will have been formed.

## Geographic Isolation

Given the great sweep of geologic time, such events as the uplift of a mountain range, the gradual increase in the size of a river, or the movement of a continental plate can result in the physical separation of a species into two or more distinct subpopulations. When the size of the barrier is large, it can effectively prevent gene flow between the subgroups.

In Figure 5.3 a small stream flows through a valley that contains a population of insects with a flight range of less than 10 feet. Because the valley is small, and the river is only a few feet wide, there is the potential

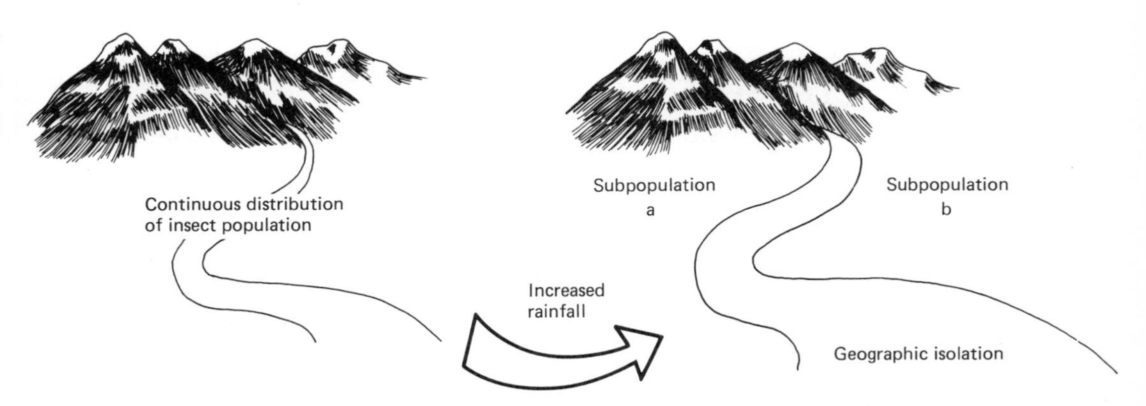


**Figure 5.3** Geographic isolation. After a period of increasing rainfall, the river widens and becomes a geographical barrier between subpopulation a and subpopulation b.

for open mating among all members of the insect population. However, the climate of the region gradually changes, and, as annual rainfall increases, the size of the stream alters from its original width of two to three feet to 15 to 20 feet. Because the insects are weak flyers, the stream now represents a geographic barrier between the two subpopulations.

The establishment of the geographic barrier causes isolation of subpopulations a and b and, even though the valley is small, events in the two gene pools will vary. Mutation is a random event and it is extremely unlikely that identical mutations will occur at the same rate in both a and b. Also, at least minor variations in environmental factors are certain to occur, so that natural selection will not be the same in both areas. The end result will be differences in gene frequencies between the gene pools at a and b. Small differences accumulate over the years, producing *divergence*—the separation of a single stock of organisms into distinct groups adapted to the specific environment in which they must survive. Divergent adaptive changes may accumulate to the point where reproductive isolation occurs; i.e., new species have been formed.

Geographic isolation has occurred in the past on a global scale as a result of continental drift. During the later stages of the Paleozoic Era, all the land masses of the Earth were united into a "supercontinent" called Pangea, which began breaking up into small continental plates about 225 million years ago. As the individual tectonic plates separated, the plant and animal populations of each plate became geographically isolated from one another. The unique fossil records of individual continents and the biogeographic distributions of many plant and animal species can be explained by this ancient geologic event.

## Spatial Isolation

The continuous distribution of a species over a broad geographic range may be as effective in producing reproductive isolation as is the establishment of a physical barrier. Variations in environmental conditions over the range of the organism may produce geographically recognizable subpopulations (*geographic races*) that may be separated from one another by considerable distances. Except for migratory animals such as birds or wide-ranging species of mammals, it is unlikely that members of geographic races will meet to interbreed, and thus gene flow between them is prevented. In time, such geographic races may diverge to the point of reproductive isolation, and new species will be formed.

Richard B. Goldschmidt studied the gypsy moth (*Lymantria dispar*) in its continuous distribution through the Japanese Islands and eastern Asia and concluded that ten geographic races had been formed as a result of

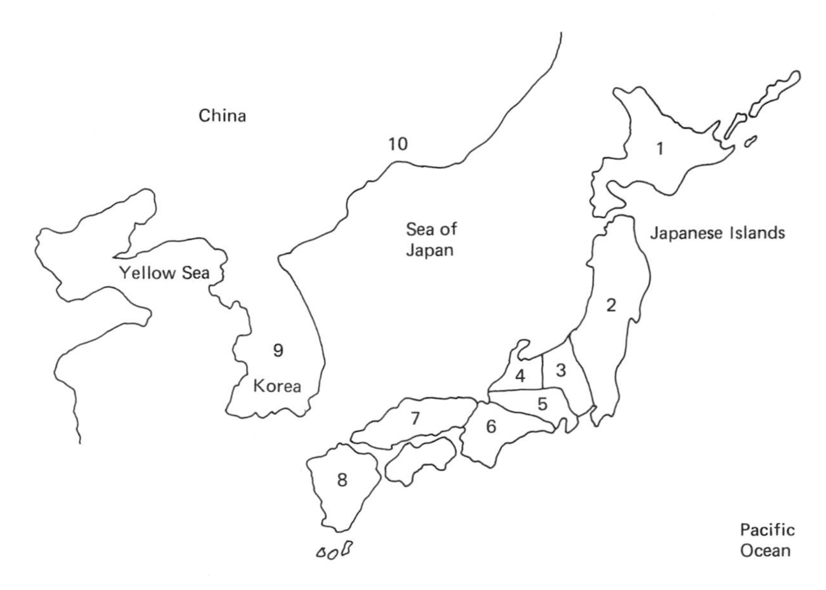


**Figure 5.4** Spatial isolation. Geographic races of the gypsy moth (1–10) have been formed because of environmental gradients that occur over the broad geographic range of the moth species.

environmental variation acting on effectively isolated gene pools of that species (Figure 5.4). These races are distinguished by phenotypic variations that reflect differences in gene frequencies within their gene pools.

One trait that illustrates the effect of environment on gene frequency is the time of hatching of eggs. The gypsy moth spends the harsh winter months in the resistant egg stage of its life cycle. The eggs hatch in the spring and the larvae begin immediately to feed on the foliage of the budding trees. If the eggs hatch too early in the spring, the larvae face the danger of starvation or of exposure to killing frost. Thus, the time of hatching is important to the survival of the species.

In its distribution throughout the Japanese Islands, the gypsy moth encounters climatic conditions that range from subarctic in the northernmost island to subtropical in the southernmost. Goldschmidt studied ten localities within this range and found that the time of hatching is progressively later from south to north. This difference in hatching time is a balance between heredity and environment. The eggs hatch in response to the warming temperatures of spring, but the genetic potential for environmental response varies from the southern races to the northern races, so that even if exposed to identical environmental conditions, there remains a heritable difference in hatching time. This genetic control prevents the early hatching of eggs of the northern race should there be an unseasonably warm day or two in late winter.

Some biologists consider the gypsy moth races to be incipient species in which the gene pools have diverged to the extent that they border on reproductive isolation. Experimental evidence supports this idea; matings between members of the individual races produce offspring that interbreed with difficulty, if at all.

## Isolating Mechanisms

Interbreeding between members of similar species is prevented by the establishment of isolating mechanisms—barriers to reproduction. Some common examples of isolating mechanisms are: mechanical isolation, seasonal isolation, behavioral isolation, hybrid sterility, hybrid inviability, and selective hybrid elimination.

*Mechanical isolation* results from structural differences in the genitalia of two closely related species that make interbreeding impossible. *Behavioral isolation* results from differences in the habits of animal species. For example, the acceptance of a mate by the female of many bird species is predicated on a complex courtship "dance" performed by the male of that species. If the proper dance (behavior pattern) is not followed, the male will not be accepted. The reproductive cycles of many plant and animal species are correlated with seasonal climatic conditions. *Seasonal isolation* occurs when species become sexually active at different times of the year.

*Hybrid sterility* has been mentioned in the example of the horse and the ass—mating occurs but the offspring is infertile because of genetic or chromosomal incompatibilities. In *hybrid inviability,* the degree of incompatibility is so great that the offspring die.

*Selective hybrid elimination* occurs when members of a species are in intense competition for a common resource—as in the example of Darwin's finches. Under such conditions, some members of the population seek out alternate resources and each subpopulation undergoes a distinct set of adaptive changes. Initially, hybrids occur between the two groups, but, as the hybrids lack specialization for either resource, they will not be able to compete against the offspring of those individuals that breed *within* subpopulations. The rate of adaptive change under these conditions may be very rapid.

## Linear Species Formation

Species formation does not always result from the divergence of a species population into two or more reproductively isolated subpopulations. At times, a new species is formed through gradual environmental changes that produce a series of adaptive changes within the gene pool of a single line of descent. Many anthropologists believe that modern man (*Homo sapiens sapiens*) has evolved from Neanderthal man (*Homo sapiens neanderthalensis*), and that Neanderthal man, in turn, evolved from *Homo erectus* (e.g., Java man and Peking man). The gradual transition from *Homo erectus* through *H. sapiens neanderthalensis* to *H. sapiens sapiens* occurred over a period of millions of years and was caused by changes in environmental conditions operating on the gene pool of the genus *Homo*. The important point is that species formation occurred in a linear rather than divergent pattern (Figure 5.5).

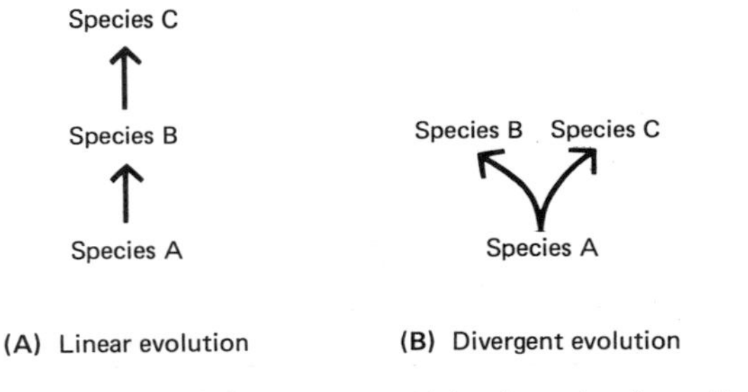


**Figure 5.5** Linear and divergence models of species formation.

## Changes in Chromosome Number

Occasionally the intricate sequence of meiotic cell division fails to function in its usually precise manner and gametes are formed that contain twice the normal number of chromosomes (Figure 5.6). When this gamete unites with a gamete containing the normal complement of chromosomes, a zygote is produced that contains three times the normal chromosome number (called a *triploid*). Triploids are generally inviable or, if they survive, are infertile. In some cases two gametes with the doubled number of chromosomes unite and form a zygote that contains four chromosomes of each kind instead of the usual two. Such individuals are called *tetraploids* and are often capable of interbreeding successfully with other tetraploids—*but not with the parent type.* Such tetraploid individuals are reproductively isolated from the original gene pool and a new species has been formed within a single generation! This type of speciation is occasionally observed in plants but is rare among animal species. Tetraploid speciation does not result from geographic isolation of gene pools nor does natural selection appear to play a role in the development of the tetraploid individual.

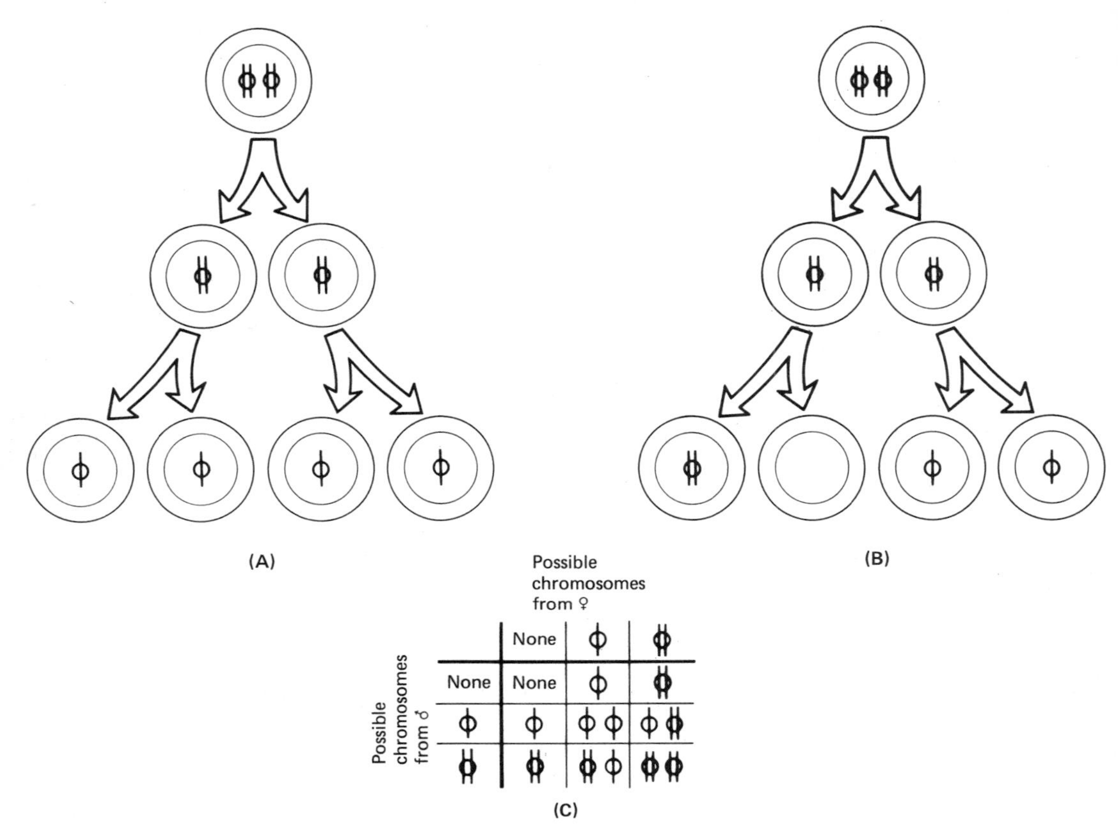


**Figure 5.6** Chromosomal anomalies. Failure of the chromosomes to distribute normally during meiotic cell division may result in gametes with an abnormal chromosome number. (A) Normal meiotic division. Each gamete has a single chromosome of the pair. (B) Meiotic division in which the two strands of one chromosome have failed to separate during the second division. This situation produces one gamete with two chromosomes of the pair and one gamete with none. (C) Possible chromosome combinations resulting from crosses of gametes with normal and abnormal chromosome numbers.

# 6 Evidences of Evolution

## Introduction

If organic evolution has occurred, there should be observable similarities among those species that have diverged from a common ancestor. Those species with the most recent common ancestor should have the most traits in common. Scientists of the nineteenth century sought evidence of such similarities in the newly developing fields of comparative anatomy, comparative embryology, paleontology, and biogeography. So far in the twentieth century, we may add to this list the studies of genetics, comparative biochemistry, and molecular genetics.

## Comparative Anatomy

The study of comparative anatomy has traditionally offered the strongest evidence in support of the theory of organic evolution, as study of the organs, tissues, and systems of related groups of animals reveals many fundamental similarities in structure. An anatomical system of special interest in such comparisons is the skeletal system. Because this is the part of the vertebrate animal most commonly fossilized, comparisons between

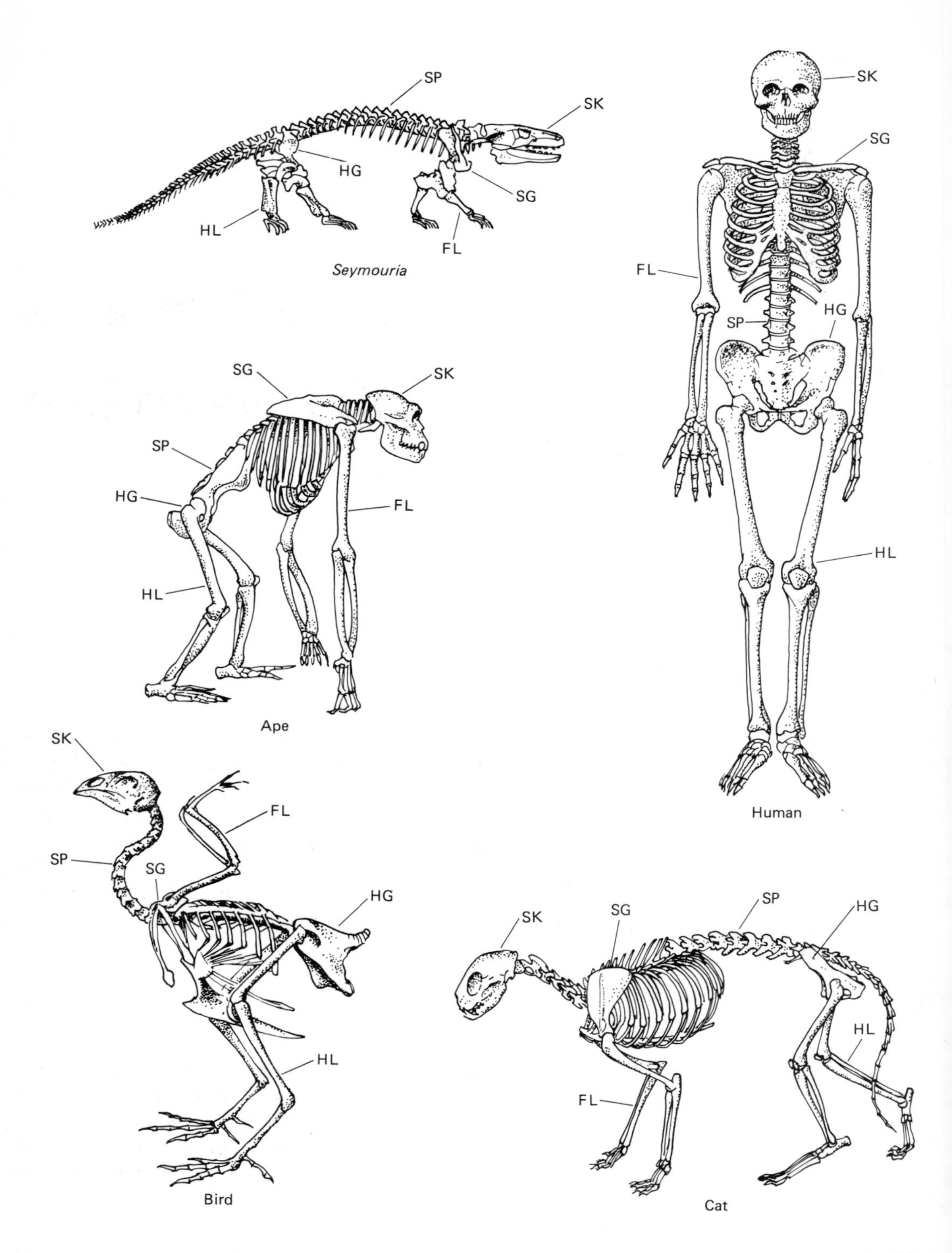
SP
SK
HG
SG
HL
FL
Seymouria
SK
SG
FL
SP
HG
HL
Human
SG
SK
SP
HG
FL
HL
Ape
SK
FL
SP
SG
HG
HL
Bird
SP
SK
SG
HG
HL
FL
Cat

living and extinct species are usually based on comparative anatomy of this system.

Figure 6.1 shows the skeletal system of five vertebrate animals. It is obvious that posture and general appearance differ among these skeletons, but it is also apparent that all of the skeletons are constructed along a common ground plan: skull, spinal column, shoulder girdle, hip girdle, and the limbs. The degree of similarity is all the more striking when we realize that one of these animals, *Seymouria*, has been extinct for over 230 million years.

The observation that fundamental similarities underlie the more apparent differences among groups of animals is explained on the basis of common ancestry—in modern terms, descent from a common gene pool. Modern species of reptiles, birds, and mammals have evolved from primitive amphibian ancestors very similar to *Seymouria* and have retained many features in common.

Equivalent structures derived from a common genetic ancestor are called *homologous* structures. The bone structure of the vertebrate forelimb provides an example of homology. Each of the forelimbs in Figure 6.2 is adapted to a particular function: *Seymouria's* for walking; the mole's for digging; man's for grasping and manipulation of objects; and the bird's for

**Figure 6.2** Homologous structures in vertebrate forelimbs. H– humerus; U– ulna; R– radius; C– carpels; MC– metacarpels; P– phalanges.

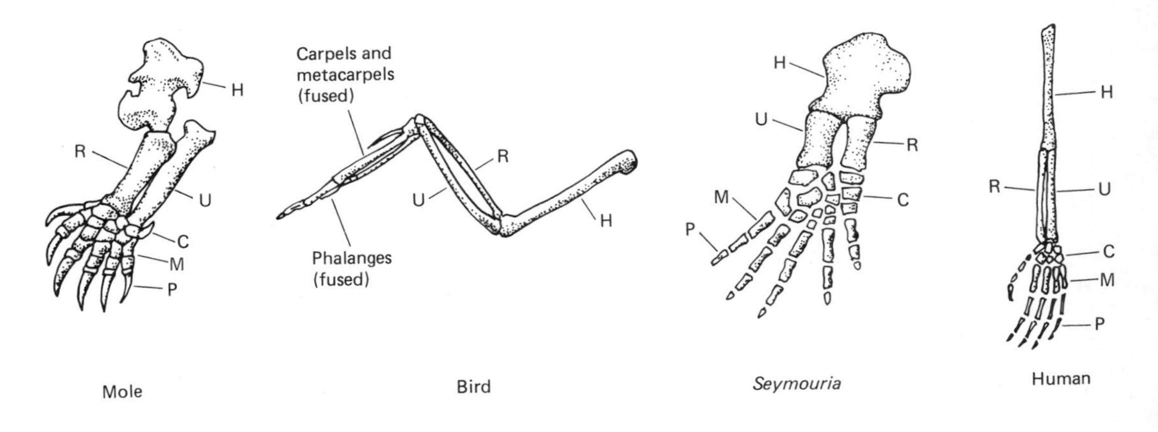


**Figure 6.1** Comparative skeletal anatomy of vertebrates. SK– skull; SG– shoulder girdle; SP– spine; HG– hip girdle; FL– forelimb; HL– hind limb. The two girdles are composed of several bones and are important because they serve to connect the limbs to the axial skeleton and provide both support and freedom of movement for the appendages.

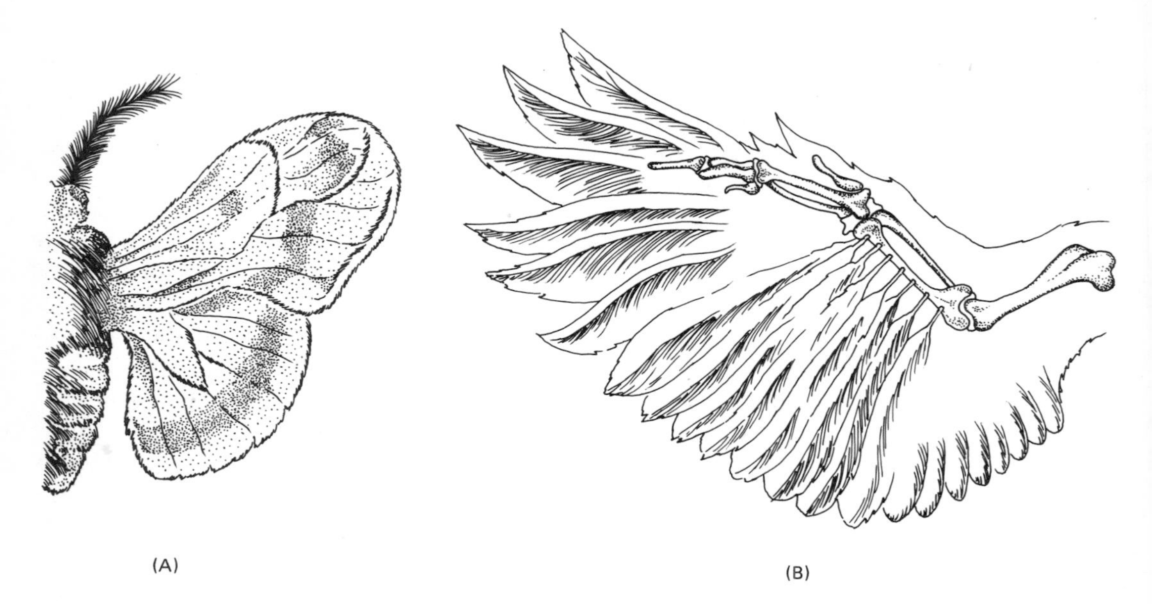


**Figure 6.3** Analogous structures. The wing of an insect (A) and the wing of a bird (B).

flying. These adaptations to life-style are reflected in the shape and relative lengths of the bones. However, despite the wide range of functional adaptation, precisely the same bones occur in each forelimb.

Another observation of comparative anatomy is that similar or parallel adaptations occur in groups of organisms with a very remote common ancestry. Birds and insects have wings for flight, but the lines of descent leading to each of these modern forms diverged far back in geologic time. A comparison of the anatomy of the wings of the two groups reveals that the similarities are superficial rather than structural (Figure 6.3). The skeletal system of the insect is external as opposed to the internal skeleton of the bird. In the bird wing, skin and feathers are stretched over the framework of the internal skeleton, which provides both strength for the wing and attachment for the muscles that activate the wing. In contrast to the anatomical arrangement of the bird wing, the beetles have two pairs of wings, one of which is a hardened membrane that covers and protects the more delicate flying wing. The membrane of the flying wing is strengthened by tubular veins rather than an internal skeleton and the muscles that activate the insect wing are largely contained within the thoracic region of the insect's body.

Clearly, the wings of birds and insects have arisen independently of one another, the result of different mutations that occurred in separate lines of descent. Anatomical structures that perform the same function but which are structurally unrelated are called *analogous* structures.

## Comparative Embryology

All vertebrate animals begin life as a fertilized egg (zygote). The zygote produces a new individual of the species through cell division, differentiation, and specialization. Each event of this sequence is mediated by the genetic potential inherited from the parents. Comparison of vertebrate embryos at equivalent stages of development reveals strong resemblances among the embryos during their early formation (Figure 6.4). As the embryos mature, these early similarities are lost, and traits more typical of the adult of the species are produced. At eight weeks the human embryo has gill slits, a rudimentary tail, and a circulatory pattern more homolo-

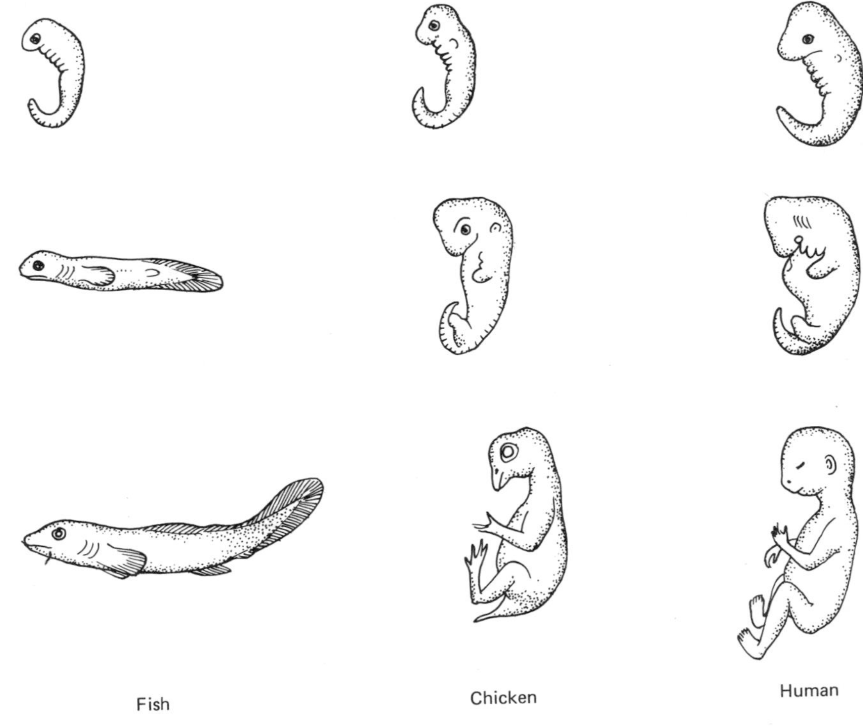


**Figure 6.4** Comparative embryology. Embryos of a fish, chicken, and human are presented at comparable stages of development.

gous to that of the fish than the adult human—traits that are lost or greatly modified during the latter stages of embryonic development.

This type of observation produced the theory of *recapitulation*—the idea that the developmental stages of an organism repeat events of its evolutionary history. For example, the toad begins life as a fertilized egg, becomes a fishlike tadpole and breathes with gills, and finally becomes a terrestrial air-breathing adult. Since amphibians are thought to have evolved from fish, this sequence of events repeats a part of the evolutionary history of the toad.

While a strict application of the theory of recapitulation to all events of embryonic development is not possible, it does seem likely that similarities in vertebrate development reflect the retention of genetic material from an ancestral line of descent. The disappearance of such "primitive" traits during maturation is explained by the activity of genes that have entered the gene pool through mutation and have been retained because they confer an adaptive advantage to the adult of the species. These genes modify the expression of the "ancestral genes" expressed in the early embryonic stages.

## Paleontology

The fossil history of life provides several arguments in favor of the theory of organic evolution: (1) extinction of species, (2) evidence of progressive change, and (3) fossil forms whose characteristics are transitional between major groups of organisms.

The occurrence of extinct species in the fossil record emphasizes the fact that variation has always been a part of life and that species are not "fixed." Life has existed in many forms throughout the history of the Earth and the fossil remains of modern forms appear only recently, within the last 60 to 70 million years.

The fossil record provides evidence that changes in living organisms have taken place slowly, through the gradual accumulation of adaptive traits. The fossil record of the horse is a good example of such *progressive* change (Figure 6.5). A series of fossil remains covering a time span of some 60 million years illustrates the steps in the transition from the dog-sized *Eohippus* of the Eocene Epoch to the larger, more specialized *Equus* of today. Specific changes include: an overall increase in size; increase in limb length in proportion to body size; fusion of the limb bones for strength in running; fusion of the separate toes of *Eohippus* into the hoof of the modern horse, also for strength; and specialization of the teeth for

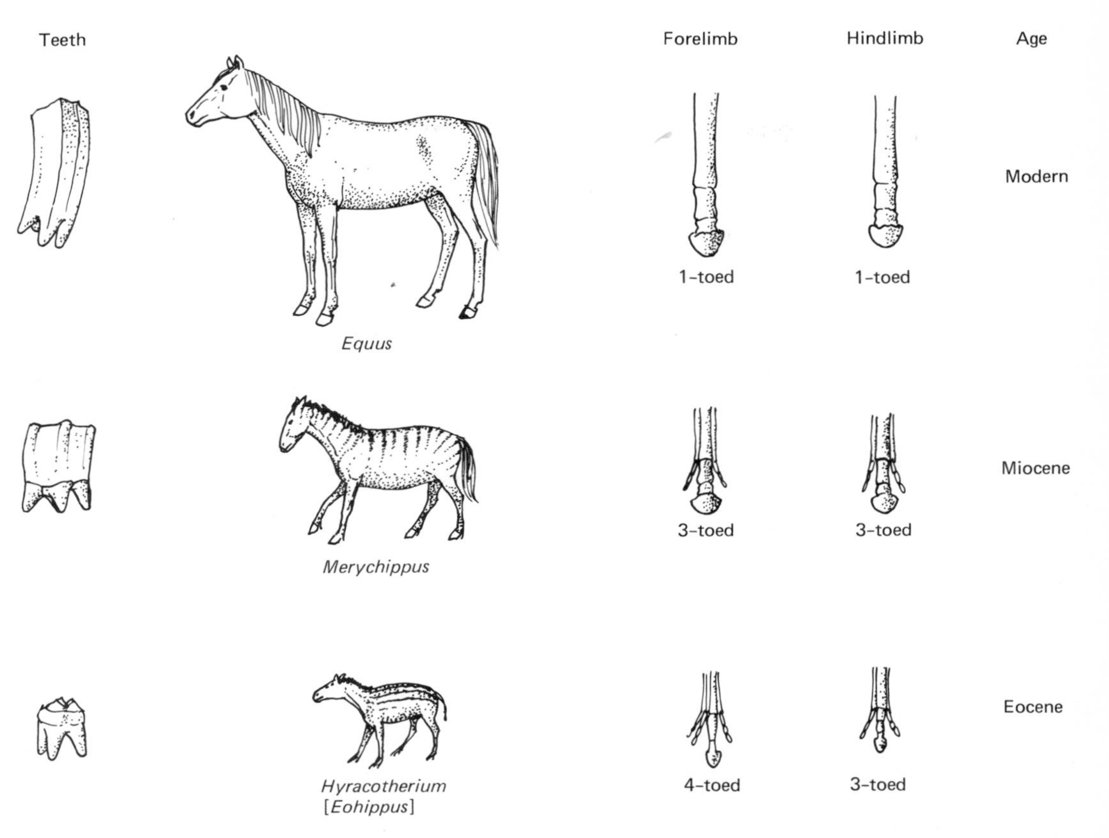


**Figure 6.5** Stages in the evolution of the horse. The progressive nature of evolutionary change is illustrated in (1) the elongation of the teeth, (2) the elongation, fusion, and reduction in number of limb bones, and (3) the overall increase in body size.

eating grasses. The distinct steps in the progressive change of these traits are evidence that an evolutionary process has occurred.

Finally, the fossil record contains extinct forms that are transitional between lines of descent. *Archeopteryx* lived during the Jurassic Period of the Mesozoic Era and the fossil remains of this animal display the teeth and tail of the reptile, the wings and feathers of the bird, and many skeletal details that are intermediate between the reptilian and avian classes (Figure 6.6). Likewise, the fossil record of the therapsid reptiles of the late Permian and Triassic Periods show the gradual transition of this reptilian line into the first mammals, which appeared during the Jurassic Period. Such transitional forms provide observational evidence that life-forms have arisen through descent with modification from preexisting forms.

**Figure 6.6** *Archeopteryx.* A fossil animal considered to be transitional between reptiles and birds.

# Biogeography

Observations of the distribution of living and extinct species created doubt in Darwin's mind concerning special creation. He believed that distribution patterns of the world's flora and fauna are best explained by the influence of climatic and geographic factors on the evolution of species. The history of the genus *Magnolia* provides a good example of environmental control over the distribution of a species.

In recent times the *natural* distribution of magnolias has been limited to two widely separated geographic areas—the southeastern United States and eastern Asia. However, paleontological evidence shows that in the past magnolias had a continuous distribution over the entire geographic region that now separates these two isolated populations. The broader distribution of the genus *Magnolia* occurred during a warm period of Earth history. With the coming of the ice age, a cooling climate gradually eliminated the species of *Magnolia* from all of its range except the relict areas it still occupies. Following the glacial period, world climate warmed again and increased the potential range of magnolias, as reflected by the success with which members of this genus have been transplanted as ornamentals over much of the United States.

In addition to changes in world climate, such geologic events as mountain building and continental drift have also influenced the geographic distribution of species by isolating populations of plants and animals and by preventing dispersion and interbreeding. The physical movement of the continental tectonic plates has at times resulted in the migration of plates into new climatic zones, thus causing changes in the evolutionary history of the flora and fauna present on that continent.

## Comparative Biochemistry

In recent years the study of the molecules that make up living systems has provided new evidence in support of the monophyletic model of organic evolution. Biochemical organization preceded structural complexity in the history of life, and the genes controlling the processes of heredity and respiration are as old as life itself (see Chapter 8). In modern organisms there are great similarities in the chemicals involved in these two processes. The gene is chemically the same wherever it occurs and only variations in the base sequence of DNA spell out differences between the earthworm and the whale. Adenosine triphosphate (ATP) is the common currency of respiration in every living cell and there are many similarities among the respiratory enzymes of all living organisms. The existence of such fundamental chemical homologies is taken as evidence that all living organisms can ultimately be traced back to a common genetic stock and that the genes controlling the most basic biochemical processes of living systems have been stabilized in the gene pools of untold numbers of species of organisms through the millions of years life has been present on Earth.

## Mutations and Phylogenetic Diversity

The understanding of gene function and gene mutation at the molecular level provides new insights into the study of organic evolution. Since genes code the production of proteins, analysis of the amino acid sequence of similar proteins found in individuals from two species can be used as a technique for determining closeness of evolutionary relationship. For example, cytochrome enzymes are found in all organisms. By comparing the amino acids of the cytochromes of different species, the number of similarities (and differences) can be noted. Groups of organisms in which the cytochromes are very similar are interpreted as having a close similarity in genetic makeup which in turn indicates that the two organisms have a recent common genetic ancestor. Conversely, species that have greater amino acid diversity (thus, greater genetic diversity) represent lines of descent that have diverged farther back in geologic time.

By studying the protein composition of organisms, molecular biologists can prepare "family trees" based on degrees of genetic similarity and compare these with "family trees" based on evidence from the fossil record. In general, the results of the two methods of establishing evolutionary relationships agree very closely. Such confirmation of scientific interpretation from two widely divergent approaches lends great strength to the theory of organic evolution.

# 7 Evidences of Natural Selection

## Introduction

Up to this point we have discussed Darwin's concept of natural selection and have discovered that the gene theory of inheritance provides the raw material for the selection process and that population genetics establishes the dynamics of the interaction between the gene pool and environmental selective factors. In the present chapter we will consider two examples that clearly illustrate that natural selection does occur in natural populations. We will also consider an experiment designed to determine the *source* of favorable mutations within a gene pool.

In order to determine whether natural selection has occurred in a population, Raymond Pearl, a biologist at Johns Hopkins University, proposed the following criteria: (1) that there be observable variation among members of the population, (2) that there be differential survival among the variant phenotypes, and (3) that this differential survival lead to differential reproduction. The ultimate result of differential reproduction is change in the frequencies of those genes that control the trait undergoing selection. The following examples illustrate that these conditions have been met in nature.

# Insecticide Resistance

The time factor is often stressed in the discussion of organic evolution. In general, environmental changes occur slowly and corresponding phenotypic changes in populations accumulate at a slow rate. However, the activities of man have sometimes resulted in an abrupt and extreme selection pressure being applied to certain animal populations. Some species, not able to adapt to such extreme situations, have become extinct. The carrier pigeon, unable to avoid intensive hunting by man, was literally exterminated. The pileated woodpecker is thought to be nearly extinct as a result of a more subtle but equally effective environmental stress—the inexorable reduction of suitable habitat. Other species have shown a remarkable ability to adapt to a changing environment; rats and English house sparrows thrive in the "man-made" environment of urban areas.

A dramatic model for the demonstration of adaptability to a wide range of environmental conditions is provided by the class Insecta. Since the time of their origin in the Devonian Period (approximately 400 million years ago) insects have proliferated evolutionarily until they now number over one million species. Insect species occupy almost every available habitat—from lakes to deserts, from below ground to high mountain summits. Some of the specific adaptations that account for the success of the insects are: (1) their generally small size and modest individual requirements for food and water; (2) an exoskeleton that provides mechanical protection and retards desiccation under arid conditions; (3) some species undergo metamorphosis, which provides for a variation in habitat during the life cycle, or they have dormant stages that enable them to survive periods of stress (such as winter or a dry season); (4) the ability to fly, which provides for an extended territorial range and the opportunity to escape local environmental stresses; and (5) a tremendous *biotic potential* that provides for a large population and also guarantees that there will be great genetic diversity expressed through genetic recombination.

The very traits that make insects so successful also magnify the problems man faces when attempting to control those insect species detrimental to him and his activities. In all, only about one thousand species of insects are considered as serious pests, yet this relatively small number can exact a heavy toll in agricultural depredation and property damage. Also, insects cause a great deal of human suffering and death as vectors of such diseases as malaria, yellow fever, and typhus. Because of the harmful activities of some insects, man has attempted for many centuries to find an effective means to reduce their numbers. Traditional weapons in insect control have been the drainage of swamps, screening of doors and windows, and other practices of elementary hygiene. More recently, the use of chemical insecticides has given hope that insect pests might at last

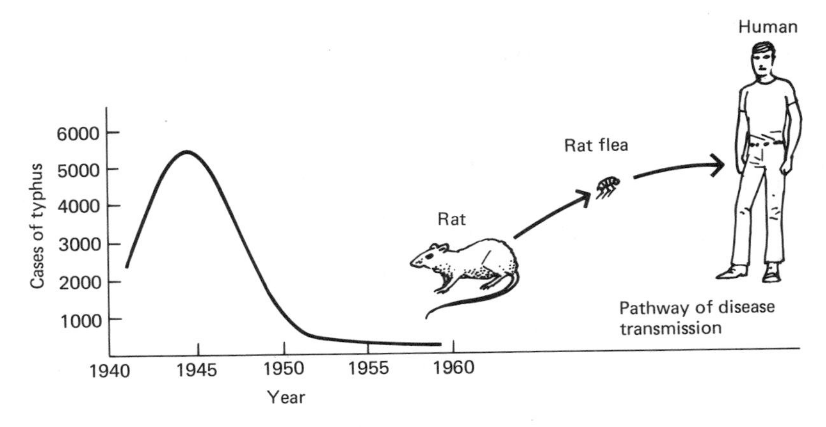


**Figure 7.1** Reduction of typhus as related to DDT usage. Typhus is vectored by the rat flea. The use of DDT to control insects during the 1940s resulted in a dramatic reduction in typhus cases in areas where normal sanitation had been disrupted by World War II. (After Benarde, Melvin A., *Our Precarious Habitat.* W.W. Norton & Co., Inc., N.Y., 1973.)

be effectively held in check. However, one characteristic of early insecticide chemicals was that their residues persisted in the environment for only a brief period. Insects in protected places might emerge after a short time and carry on their activities undisturbed, or the treated area could be repopulated by immigration from surrounding, unsprayed areas.

With the development of such persistent insecticides as DDT (dichlorodiphenyl-trichloroethane) it was thought that more complete insect control could be obtained. DDT remains on foliage and in the soil for a long time and any insects that might have avoided the original spray or immigrated into the area after the spray period would still be exposed to the toxin. Initial results with DDT were every bit as impressive as predicted. Figure 7.1 illustrates the type of benefit for human populations that can be derived from insect vector control. This figure shows the reduction in typhus incidence resulting from control of the rat flea that transmits the typhus organism from rat to man. Comparable results from DDT control programs were obtained with such pests as flies and mosquitos. Their hopes buoyed by these dramatic results, some biologists were predicting total eradication of major insect pests.

Despite the impressive early results obtained with DDT, by the late 1940s some disturbing news was reported. In Italy it was discovered that flies were no longer killed by DDT sprays because the flies had developed a genetic immunity to the insecticide. Similar reports came from other countries until, today, insecticide resistance has been reported for over three dozen insect species in countries all over the world.

The introduction and widespread use of DDT represents an extreme case of natural selection. The selecting agent is clearly identifiable as the chemical DDT. The mortality caused by spraying this insecticide on previously unexposed populations of insects probably exceeds ninety percent. However, the data reported above indicate that a genetic resistance is present in many insect species. The survivors of the initial DDT exposure possess a recognizable adaptive trait not present in the rest of the population and these survivors represent the breeding population from which subsequent generations will be derived. The great biotic potential of insects provides a special advantage under such extreme selection pressure for two reasons: (1) the large number of offspring greatly increases the probability of a resistant genotype occurring, and (2) the reestablishment of a large population can occur in a short time from a small number of resistant survivors.

Genetically, DDT resistance seems to be polygenic—that is, a trait that is controlled by *more than one* gene pair. Polygenic traits show a wider range of expression than simple dominant–recessive relationships. Thus, resistance to the insecticide is not an all-or-nothing trait. A populational distribution of DDT resistance is illustrated in Figure 7.2A in which survi-

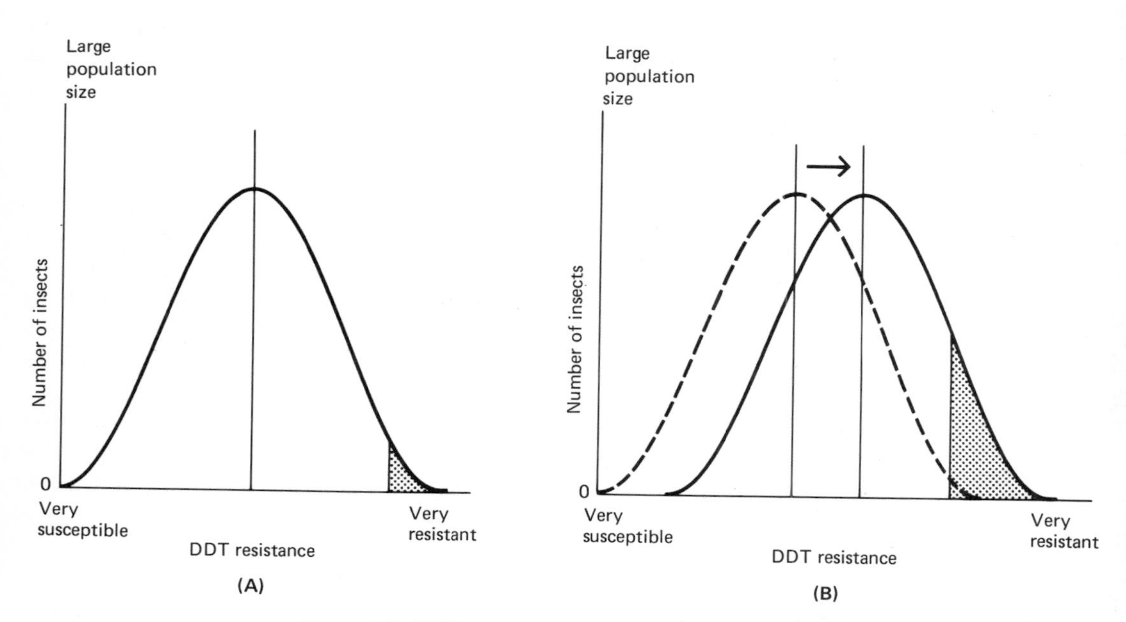


**Figure 7.2** DDT resistance and the directional effect of natural selection. (A) Resistance to DDT varies within the population. Upon exposure to DDT, only the resistant members of the population survive (shaded area). (B) Because survivors from A represent 100% of the breeding population, a greater percentage of B will be resistant to DDT. Thus, DDT is the selecting agent in natural selection, causing a change in gene frequencies between A and B (directional selection).

vors of this original population are at the upper end of the curve. Due to gamete segregation and gene recombination *not all* progeny of this breeding population are resistant. However, continued exposure to the insecticide eliminates all individuals susceptible to the toxin and, after a time, the distribution of genotypes is represented by the solid line in Figure 7.2B. The dotted line in Figure 7.2B represents the original genotype distribution.

Figure 7.2 suggests that genetic resistance was present in the original population but was not expressed until a selective agent was introduced into the environment. That is, the mutation conferring DDT resistance had occurred *before* the population was exposed to DDT. Upon exposure to DDT there was a shift in genotypes with a concomitant change in gene frequencies. In other words, the selecting agent (DDT) *caused* the change in gene frequency. Thus natural selection provides *direction* for the process of organic evolution. Without natural selection, genetic variation might occur in a population but would not result in the *directed* change in gene frequencies that we find in insecticide resistance.

Following the directed change in gene frequencies that resulted from exposure to DDT, future generations will maintain the genotype distribution of Figure 7.2B as long as DDT is present in the environment. This phenomenon is known as *stabilizing selection:* within any set of *unchanging* environmental conditions, natural selection tends to normalize a population rather than create change.

## Industrial Melanism

A classic example of natural selection operating as a result of human activities is *industrial melanism*—the phenomenon by which various species of moths have undergone genetic changes in body color in response to the darkening of their natural habitat. This environmental change has resulted from the production of particulate air pollutants, such as soot, associated with the Industrial Revolution. The phenomenon is widespread and has been reported for at least 70 species of moths in England and over 100 moth species in the Pittsburgh, Pennsylvania, region of the United States.

One of the earliest examples of industrial melanism concerned the peppered moth of England (*Biston betularia*). Normally, the peppered moth is light in color with a darker spotted pattern. In its natural habitat, the moth rests with wings spread on lichen-covered tree trunks. The color pattern of the moth confers a natural camouflage that helps it avoid detection by predatory birds (Figure 7.3). Examination of insect collections made before the year 1848 indicates that the species consisted then of

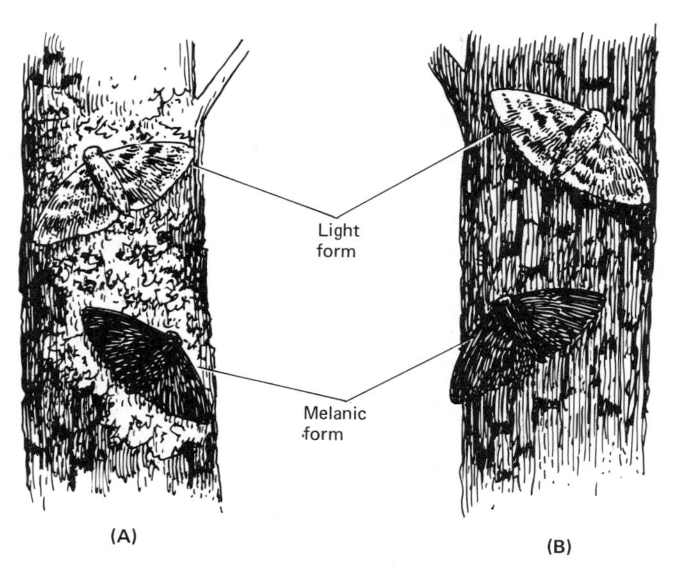


**Figure 7.3** Industrial melanism. (A) The light-colored and melanic forms of the peppered moth on an unpolluted, lichen-covered tree trunk. (B) The light-colored and melanic forms of the peppered moth on a soot-darkened tree trunk.

uniformly light-colored moths, although an occasional dark-colored (melanic) form of the moth was known to occur. Against light-colored tree trunks, the melanic form would be clearly visible to predators. During the last half of the nineteenth century, with the coming of the Industrial Revolution and the burning of large quantities of coal, the vegetation in industrial areas underwent a progressive darkening. Under these changing environmental conditions, the selective advantage of the light-colored moths was lost and the melanic form enjoyed the protection of camouflage.

In 1850, the melanic form of the peppered moth represented no more than one percent of the population. By 1900, just 50 years later, the melanic form comprised over 90 percent of the peppered moth population in the industrialized Manchester region of England. In nonindustrial regions of England the peppered moth populations retained the original high ratio of light-colored to melanic forms. This situation provides a clear example of the effects of environmental variation within a geographic region leading to observable differences in phenotype distribution. It is significant that in areas where there has been a reduction in air pollution levels the tree trunks have become progressively lighter and the light-colored form of the moth has reestablished its numerical dominance over the melanic form.

H. B. D. Kettlewell demonstrated experimentally that predation was the selective factor that caused observed changes in populational phenotype and thus changes in gene frequency. He reared large numbers of light-colored and melanic forms of the peppered moth and released them in both smoke-darkened and natural areas. When Kettlewell recaptured the surviving moths and compared recapture data of each color form from each environmental situation, he found that the recapture rate from the nonindustrial regions ran two-to-one in favor of the light-colored form; the recapture rate from the industrial regions ran two-to-one in favor of the melanic form (Table 7.1). Thus, experimental evidence supported field observation. In the smoke-darkened region, the once rare melanic form enjoyed a selective advantage over the original light-colored form of the moth.

**TABLE 7.1 Recapture Data for Light-Colored and Melanic Moths**

| | COLOR | NO. RELEASED | NO. RECAPTURED | % RECAPTURED |
|---|---|---|---|---|
| Industrial (Birmingham) | Light | 64 | 16 | 25.0 |
| | Melanic | 154 | 82 | 52.3 |
| Nonindustrial (Dorset) | Light | 496 | 62 | 12.5 |
| | Melanic | 473 | 30 | 6.3 |

Genetically, color in the peppered moth may be considered to be controlled by a single pair of gene alleles. The melanic form is produced by a dominant gene that is the mutant form of the gene for light coloration. The rapid spread of the melanic trait through the moth population can be partially explained by the fact that the adaptive mutant form of the gene is dominant to the original gene. A recessive mutant that conferred an adaptive advantage would doubtlessly have taken longer to proceed through the population.

## Source of Mutation

A situation comparable to DDT resistance among insect species is found in the resistance of bacterial populations to antibiotics such as penicillin and streptomycin. Clinical and experimental evidence clearly demonstrates that many infectious bacteria have developed a heritable resistance to antibiotics that can be traced to gene mutation. Such situations present a question of fundamental concern to biologists. What is the *cause* of favorable mutations that confer an adaptive advantage to certain individuals within a population? In the examples of DDT resistance and industrial melanism the favorable trait seemed to appear suddenly at just the proper time to aid in the survival of the species. The suggestion was offered that such gene mutations occur randomly and are simply undetectable under "normal" environmental conditions. When the proper environmental stress is introduced, those individuals having the adaptive trait represent the surviving and reproducing segment of the population. An alternative explanation proposes that the environment causes favorable gene mutations.

The idea that the environment induces adaptive change was formally proposed by the French zoologist Jean Pierre Baptiste de Lamarck in 1802. Lamarck believed that evolution occurs through the acquisition of favorable traits in response to environmental changes. Lamarck's theory provided a deceptively simple and appealing explanation for evolutionary change. For instance, the ancestors of the giraffe, undergoing severe competition for low-growing foliage, reached into the trees to obtain food. As a longer neck is advantageous in this type of feeding, through years of continual stretching to reach the high foliage, the neck of the giraffe grew progressively longer. Cause and effect are present in this explanation as are the obvious results of the process—modern giraffes have much longer necks than had their fossilized progenitors.

Darwin and modern evolutionists have opposed Lamarck's theory, although Darwin, without benefit of the gene theory of inheritance, could not discard it entirely. The alternative to the theory of acquired characteristics is that adaptive variation occurs independently of the environment and may be selected by the environment, if conditions are favorable. The giraffe has a long neck because, among its ancestors, there were individuals in the population who had the genetic potential for a longer than average neck. The adaptive advantage of a long neck might derive from either long-necked individuals being better able to obtain food, and/or their being better able to see predators. Over the millennia, continual selection for this genetic trait produced an animal with the genetic potential for an exceptionally long neck. This is another example of directional selection.

While it is true that modern biologists believe that spontaneous gene mutation is entirely random, it is necessary to provide some experimental evidence in support of this belief. Joshua Lederberg devised such an experiment.

## The Lederberg Experiment

The Lederberg experiment was designed to determine whether the gene mutation conferring antibiotic resistance occurred before or after expo-

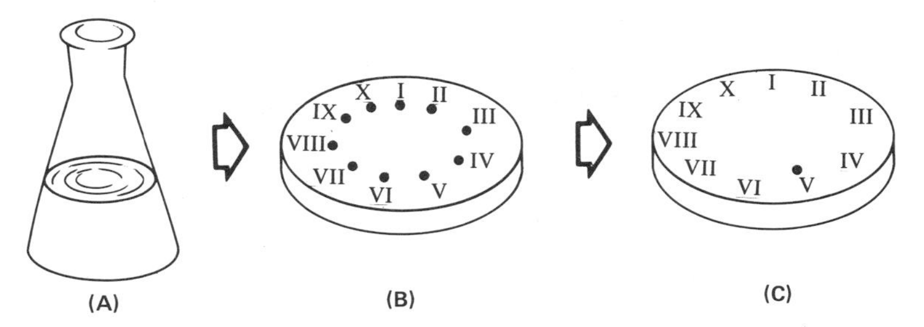


**Figure 7.4** The Lederberg experiment. (A) Parent culture of bacterial cells *not* resistant to streptomycin. (B) Single cells are taken from culture *A* to initiate populations I through X on culture medium which *does not* contain streptomycin. (C) Cells are transferred from populations I through X in *B* to a culture medium that contains streptomycin. Growth occurs only in population C.V.

sure of bacterial cells to streptomycin. From a bacterial strain lacking antibiotic resistance, Lederberg selected a *single* cell to produce a parent culture for propagation of subpopulations (Figure 7.4 A & B). He transferred cells from each subpopulation of 7.4B to streptomycin-based culture and found that growth occurred only in cells transferred from subpopulation B.V. Several repetitions of this last step gave identical results.

These results convinced Lederberg that the mutation conferring antibiotic resistance occurred in a single cell of the parent colony (B.V.) and that this cell was selected by chance to initiate subpopulation C.V. Thus antibiotic resistance was present *before* exposure of the bacterial cells to streptomycin. If the antibiotic had *caused* the favorable mutation, resistance would have appeared in many other colonies and would not have been restricted to a single line of descent.

## Summary

The above examples meet Pearl's criteria for the demonstration of natural selection: (1) there are observable variations in phenotypes; (2) there *is* differential survival of phenotypes, and (3) differential survival leads to differential reproduction. Having provided examples of the occurrence of natural selection, certain points should be emphasized: First, the result of natural selection is a change in gene frequencies, and this is a direct result of differential reproduction. Before habitat darkening by air pollution, the recessive gene conferring light color to the moth had the highest frequency in the gene pool. Under the polluted conditions that came with industrialization, the dominant melanic form increased rapidly. Kettlewell provided experimental evidence that this change in gene frequencies was a result of predation (natural selection). Second, mutation is an important factor in natural selection. The Lederberg experiment provides evidence that mutations occur randomly, rather than as a result of environmental change. The mutant gene that confers antibiotic resistance has no survival value in the natural environment of the bacterium. However, when exposed to streptomycin, the mutant gene has a positive survival value. Thus, a mutation that is neutral under one set of environmental conditions may prove beneficial under changed conditions.

Gene recombination and mutation are the sources of variation for natural selection. Natural selection operates on this variation and provides direction for the process of organic evolution. However, under unchanging environmental conditions, natural selection results in a decrease in a population's variability, a condition known as stabilizing selection.

# 8 Theories of Origin

## Origin of the Universe

Looking out from the Earth we can see the sun, the moon, a few planets, and many stars of the night sky. Stars are visible because they radiate tremendous amounts of energy into space—the product of the fusion of hydrogen atoms in their super-hot interiors. We can see the moon and planets because they reflect the light of the sun, a true star. In all, our *naked-eye* observation of the universe is limited to a few thousand stars, five of the nine planets of our solar system, and one planetary satellite, the moon. With a small magnifying lens thousands more stars become visible, and with a powerful telescope, such as the Palomar 200-inch reflector, literally millions of stars can be seen. Astronomers have observed that these stars are not spread randomly through space but are arranged in distinct gravitational systems called galaxies, each galaxy containing millions of stars.

Careful examination of the galaxies shows that they constantly move and change. Stars in some regions of a galaxy are "burning out," while in other parts stars are forming from great clouds of hydrogen gas that are shrinking under the combined gravitational attraction of the billions of atoms of which they are composed. As the "dust cloud" contracts, the crowded atoms release heat energy, ultimately producing temperatures high enough to initiate the fusion reaction that marks the origin of a star. This cycle of "birth and death" of stars is perpetuated in part by the explosion of many "burned out" stars, an event that hurls the gaseous substance of the star into space where it can once again join in the process of star formation.

Evidence for a cycle of regeneration in the universe leads to the question of the origin of time and matter. Has this cycle always been present—or is there a specific time when it began? The question of the origin of the universe has been argued by theologians, philosophers, and scientists for hundreds of years and the debate is fueled by the fact that there is no direct observational evidence concerning the events of the origin—if there was one. Current astronomic theory is based upon observation of the universe as it exists today and upon theoretical calculations of what the universe must have been like in the past—assuming uniformity of physical and chemical events in time and space. Based on these considerations, modern astronomers believe that there was a time when the universe, as we know it, began.

Astronomers date the origin of the universe to an event that took place some 15–20 billion years ago. At that time, all the matter and energy of the universe were concentrated in a gigantic ball called the *cosmic egg*. This great assemblage of matter and energy was extremely unstable and exploded, pushing swirling masses of atomized matter out in all directions. These atoms became organized into gigantic clouds of hydrogen gas large enough to form entire galaxies. Within the galactic dust clouds, smaller accretions of atoms produced individual stars and planets. Planets originate from globs of dust too small to form stars. There small spheres are the residue of stellar formation and, because of their small size, do not generate enough heat to initiate the nuclear fusion reaction.

## The Planet Earth

The Earth was formed from matter left over from the formation of the sun. Although too small to form a star, sufficient heat was generated so that the rocks of the forming Earth were in a molten state. During this stage the heavier minerals sank to the center of the sphere while the lighter minerals stratified atop this dense core. The lightest rocks solidified to form the Earth's crust (Figure 8.1).

The solid surface of the Earth was too hot to hold liquid water. The water that was to form the Earth's oceans enshrouded the planet in dense, steamy clouds—part of a stygian atmosphere far different from the envelope of gases that surrounds the Earth today. The present atmosphere is

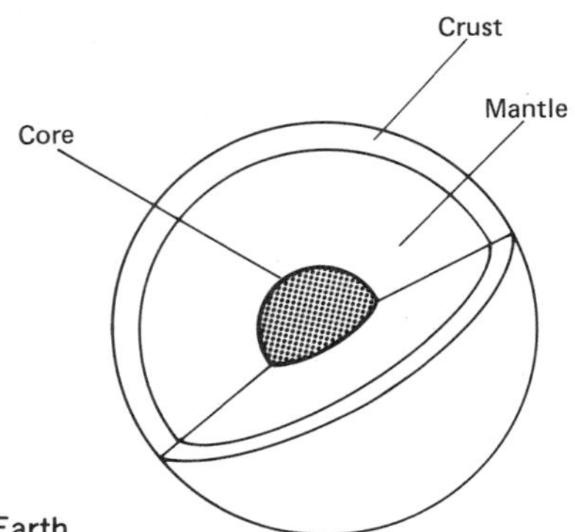


**Figure 8.1** Cross-sectional view of the Earth.

composed chiefly of oxygen and nitrogen gases, but the atmosphere of the primordial Earth contained hydrogen, methane, and ammonia (Table 8.1).

**TABLE 8.1 Components of the Earth's Atmosphere**

| AT TIME OF FORMATION | | TODAY | |
|---|---|---|---|
| Hydrogen | $H_2$ | Nitrogen | $N_2$ |
| Methane | $CH_4$ | Oxygen | $O_2$ |
| Ammonia | $NH_3$ | Carbon dioxide | $CO_2$ |

When the Earth cooled and its temperature fell to the point where liquid water could stand on the surface, a rainy period ensued lasting for thousands of years, deeply eroding the crust and producing a world ocean rich in minerals dissolved from the weathered rocks. The ocean surface was at first undisturbed by projecting landforms. The continents came later, beginning with volcanic eruptions that formed great arcs of volcanic islands comparable to the Aleutian or Hawaiian island chains. These volcanic island arcs were slowly eroded, the debris spread in thick layers over the ocean floor, and the eroded material compacted and cemented into sedimentary rocks (Figure 8.2). Subcrustal pressures compressed these sedimentary beds horizontally, causing warping and upthrust of the strata. This process resulted in the formation of great chains of fold mountains—counterparts of the Rockies and Alps of the modern world. This sequence of events produced the first continental land masses and has continued to shape the face of the Earth throughout geologic time.

By four billion years ago the Earth had taken shape—there was a world ocean, continents standing above the sea, and a stable atmosphere (though far different from our modern atmosphere). But there was no life. Even if life were present at the time of the Earth's formation, no living system could have survived the extreme heat and turbulence of the for-

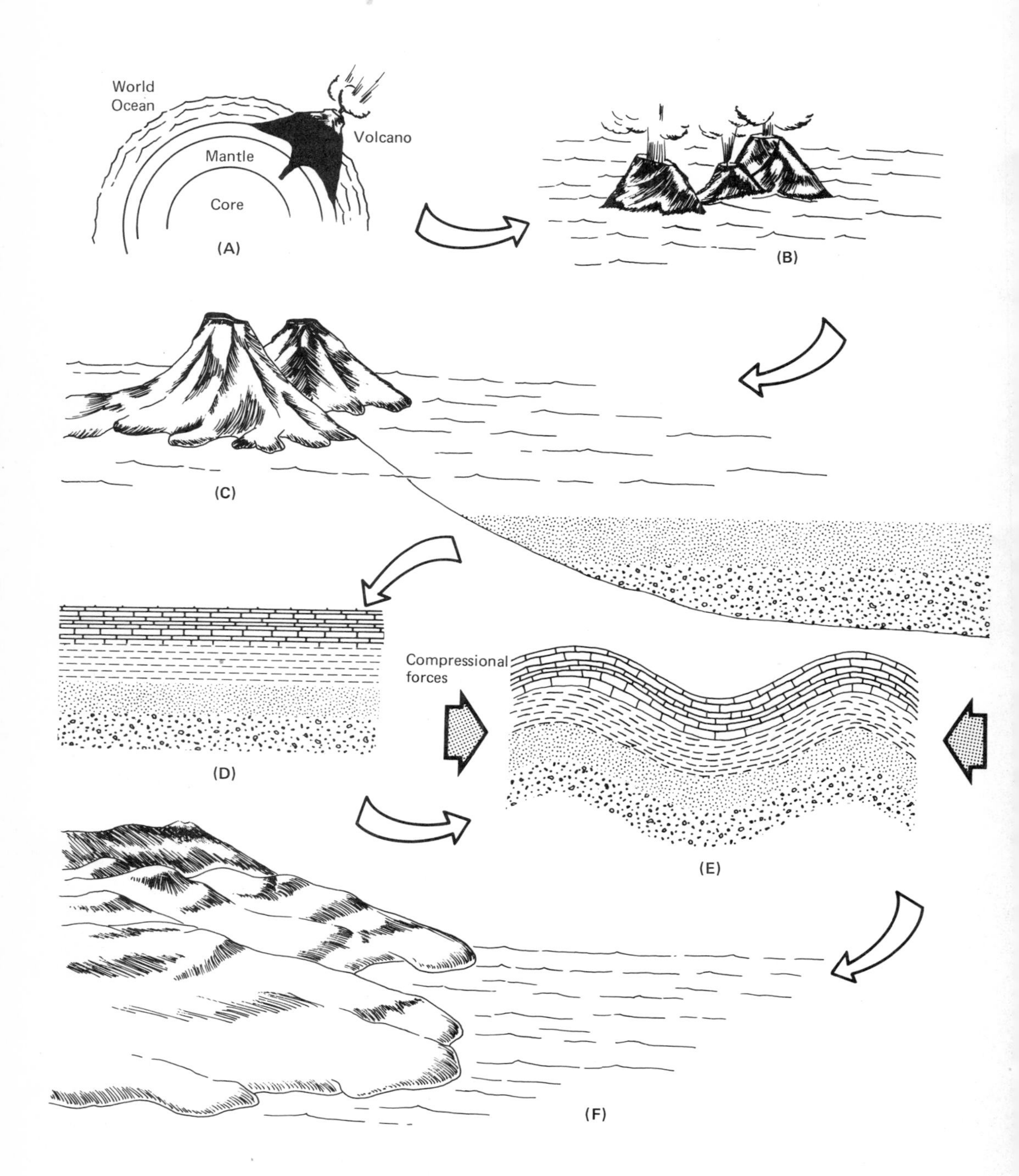
World
Ocean
Volcano
Mantle
Core
(A)
(B)
(C)
(D)
Compressional
forces
(E)
(F)

mative process. Therefore, life originated some time after the stabilization of the Earth's environment.

The earliest unequivocal fossil evidence of life appears in rock strata dated to just over three billion years. These were primitive single-celled organisms comparable to bacteria and blue-green algae. There is no evidence to indicate whether these were the earliest living forms on Earth or species evolved from some more ancient ancestor. The fact that no older fossils have been found is not conclusive evidence for either argument. The rocks of the earliest periods of Earth history have been greatly eroded, deeply buried or altered by Earth forces—processes that would have destroyed any fossil material they might have contained. Also, because of their small size and lack of hard parts, primitive single-celled organisms were rarely fossilized. For these reasons it is unlikely that there will ever be a sizable body of observational evidence bearing directly on the earliest stages of life on Earth.

Lacking such evidence, biologists have undertaken the explanation of the origin of life in much the same manner that astronomers have attempted the explanation of the origin of the universe: through the careful analysis of the available observational evidence, and a theoretical reconstruction of the conditions present on Earth at the time when life must have originated. An important consideration in this reconstruction is the observation that the structure of life becomes less complex as we proceed backward in time. Multicellular organisms first appeared about 600 million years ago, complex *eucaryotic cells* about 1.6 billion years ago and simple *procaryotic cells* date to 3.2 billion years. It is reasoned from this sequence that the very first life-forms were structurally more simple than the procaryotic cell—perhaps *acellular* living systems. The important question is, "How did these first life forms originate?" The current scientific explanation is that life originated as a result of physical and chemical events that produced a living system from nonliving matter—the theory of *inorganic* evolution.

## Inorganic Evolution

The warm shallow seas of the early Precambrian Era were rich in minerals eroded from the land and rich in methane and ammonia dissolved from

---

**Figure 8.2** Origin of land masses. (A) World ocean. (B) Volcanoes form chains of volcanic islands. (C) Sedimentary cycle produces thick layers of sediments that are formed into sedimentary rock strata. (D) Sedimentary strata. (E) Subcrustal pressures compress the sedimentary strata, causing extensive folding and uplift above sea level. (F) Continental mass. The end result of compression and uplift is the formation of folded mountain ranges. Subsequently, continental plains are formed around these centers.

the atmosphere. This chemical mixture was energized by electrical discharge (lightning) from the atmosphere and by a continual influx of high-energy ultraviolet radiation from space. This set of conditions produced a wealth of chemical reactions in the waters of the oceans, and it is hypothesized that these reactions initiated a series of physical and chemical events that culminated in the formation of life.

In 1953 Stanley Miller, a graduate student at the University of Chicago, designed an experiment that approximated the environment of early Earth. Methane; ammonia, and hydrogen gases were combined with water in an apparatus that continually recycled the mixture. An electrical discharge was maintained through the circulating brew. After seven days Miller analyzed the contents of the reaction chamber and discovered that a variety of organic molecules had been formed, including amino acids, the units of protein synthesis. Since the 1950s, Miller's experiment has been repeated and refined, with the result that most of the fundamental chemical molecules of living systems have been produced *abiotically* under controlled laboratory conditions.

The conclusion drawn from these experiments is that comparable events took place under the conditions prevailing on the primal Earth and produced a mixture of organic molecules—amino acids, DNA, ATP—and, in at least one instance, these organic precursors accumulated in a series of reactions that led to the formation of life. The criteria for life at this primitive stage are:

1. Organization. Life is the exception in the universe. It is the only state of matter that proceeds toward greater organization and complexity. Organized activity is the irreducible minimum of life. If the organization of a living cell is disrupted, death results.

2. Energy utilization. To carry out the work of life—growth, movement, metabolism, and reproduction—a living organism must have an efficient energy utilization system. This system, comparable in all living organisms, is based on the release of chemically bound energy in the process of respiration.

3. Reproduction and heredity. The maintenance of organization from generation to generation is necessary for the continuity of life. The gene is the unit of inheritance that controls this continuity in all living organisms.

The first living organism was an acellular chemical system barely distinguishable from the chemical milieu from which it had formed. The first reproduction was no more than the drifting away of a part of this system; however, each portion retained sufficient genetic material to perpetuate life and with the first separation of life into two units natural selection began its formative process. Each living system was subjected to unique environmental pressures and began the accumulation of distinct genetic variations. The process of organic evolution had begun.

# The Organization of Life

The first acellular forms gradually organized into true cells. A membrane formed around the cytoplasm, providing a measure of protection for the cell contents and regulating the movement of materials into and out of the cytoplasm. These first cells were procaryotic—lacking a nucleus and cytoplasmic specialization. However, they contained genetic material and were capable of carrying out all the chemical functions necessary for life (Figure 8.3).

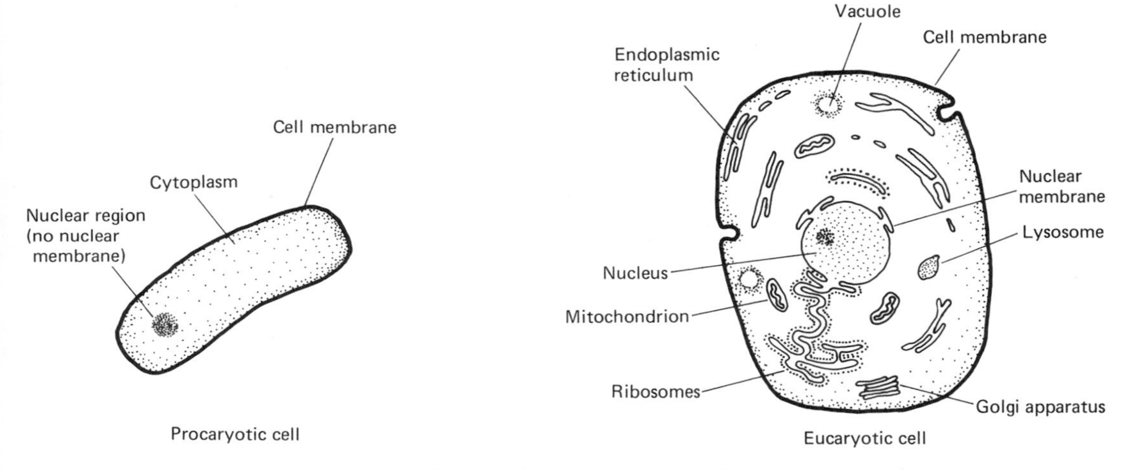


**Figure 8.3** Procaryotic and eucaryotic cells.

One limitation of the procaryotic cell is its inability to produce multicellular structures, such as the tissues and organs of higher plants and animals, and procaryotic organisms are limited to single-celled or colonial levels of organization. Colonial organisms are composed of several to many cells living together but without cellular specialization, and each cell retains an essentially independent existence (Figure 8.4).

In the eucaryotic cell the nucleus is separated from the cytoplasm by its own membrane, and the cytoplasm itself is organized into subcellular units carrying on such highly specialized functions as respiration (the mitachondrion), digestion (the lysosome), and protein synthesis (the ribosome). With the separation of the nucleus from the cytoplasm the eucaryotic cell was formed (Figure 8.3).

The great increase in internal complexity of the eucaryotic cell presents an interpretive problem in the evolution of life. Some scientists believe that the eucaryotic cell evolved from the procaryotic cell through a series of mutations which, taken together, account for its great increase in internal complexity and specialization. Others feel that the degree of

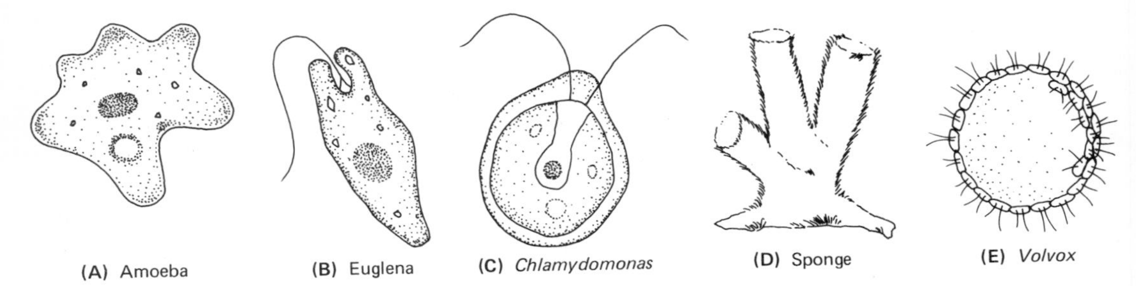


**Figure 8.4** Single-celled and colonial stages of organization. Single-celled organisms (A, B, and C) live independently and are capable of carrying on all functions of life within their single cell. Colonial organisms (D, E) are composed of many cells, but there is little or no cellular specialization and each cell remains essentially independent of the other cells in the colony.

change is too great to be explained on the basis of mutation and natural selection alone. They explain the origin of the eucaryotic cell on the basis of *symbiosis.*

Symbiosis is a biological relationship between two organisms in which there is an intimate physical contact from which both organisms benefit. For example, the lichen is a combination of an alga and a fungus growing together, the algal cell living within the fungal mycelium. The fungus benefits from the food produced by the photosynthetic alga, and the alga benefits from the water and inorganic nutrients provided by the fungus.

According to the symbiosis theory, eucaryotic cells were developed when large procaryotic cells engulfed smaller procaryotic cells. Some of these smaller cells were retained in the cytoplasm and gradually established a symbiotic relationship by becoming specialized to meet specific requirements of the larger cell—e.g., respiration, digestion or photosynthesis. As they became specialized, the smaller cells lost many of their more generalized structures and functions. The fact that mitachondria and chloroplasts contain DNA and are capable of independent division is cited as evidence that they were once free-living, independent organisms.

Whatever the origin of its internal complexity, the eucaryotic cell marks a major advance in the evolution of life—the specialization of the cytoplasm permitting the formation of multicellular organisms.

## The First Plants

The first living organisms were bacterialike, their food provided by the abiotic processes that continually formed new organic molecules in the oceans. These molecules were taken directly into the cell and used as food. However, as the number of living organisms increased, the competition for this food supply became intense and the abundance of life was controlled by the rate at which abiotic chemical reactions could replenish the food supply. One effect of such environmental stress can be the extinction of some species, but another possible result is the development of a new adaptive strategy for survival. In this instance the adaptive strategy was metabolic rather than structural.

The importance of food is twofold: (1) food molecules are used to synthesize cell parts for growth, repair, and reproduction, and (2) some food molecules are respired for energy. The energy released in respiration is incorporated in the food molecule at the time it is formed. The energy for the abiotic synthesis of organic molecules came from lightning or ultraviolet radiation—energy sources damaging to living tissue, and so the cell itself could not carry on this type of reaction. Sunlight also provides sufficient energy for the synthesis of organic molecules and is *not* harmful to most cells. The utilization of sunlight in the biosynthesis of organic molecules required the evolution of a biological system capable of capturing the sun's energy and incorporating a fraction of that energy into a chemical bond. Some procaryotic cells evolved a light-sensitive molecule called *chlorophyll,* which can absorb light energy and, along with the appropriate enzymes, use this energy to combine carbon dioxide and water in the synthesis of sugar. This process, called *photosynthesis,* freed life from its dependence on the abiotic production of food.

General Equation for Photosynthesis

$$CO_2 + H_2O \xrightarrow[\text{chlorophyll}]{\text{sunlight}} \underset{\text{(sugar)}}{\overset{\text{food molecule}}{C_6H_{12}O_6}} + H_2O + O_2$$

Photosynthetic cells were the first plants—provided with carbon dioxide, water, inorganic nutrients, and sunlight they were capable of a completely independent existence. From the time of their origin, plants have formed the basis of the ecological food chain, providing food for all other living organisms.

A by-product of photosynthesis is oxygen gas ($O_2$). As the number of photosynthetic organisms increased, the oxygen produced by photosynthesis reacted with atmospheric ammonia and methane and eventually produced the chemical balance found in our modern atmosphere. As the atmosphere gradually became rich in oxygen gas, the chemistry of life had to change to survive in the new environment. Because of its great chemical reactivity, oxygen has become an essential part of the respiratory system of nearly all living organisms.

## The Animals

The first animals evolved from colonial algae that were only marginally efficient in photosynthetic ability and had to supplement their diet by engulfing and digesting the cells of more efficient plant species. These organisms became more and more dependent on ingested food and eventually lost the ability to carry on photosynthesis altogether, thus producing the first animal species. From their beginning animals have been aggressive, motile forms, actively seeking and ingesting food.

## Kingdoms of Life

The three kingdoms of living organisms are the plants, the animals, and the protista. The protista include all organisms with procaryotic cells. Modern representatives of this group are the bacteria and blue-green algae. Because they are procaryotic organisms, organization among the protista does not progress above colonial level. Plants and animals are made up of eucaryotic cells and range from single-celled forms to the more familiar, highly complex multicellular species.

## Multicellular Forms

By the end of the Precambrian Era, life had progressed from the simple acellular systems produced by inorganic evolution, through the stage of primitive procaryotic cells, to the development of the eucaryotic cell. With the advent of the eucaryotic cell a new adaptive strategy was available to life—the formation of the multicellular organism. The oldest multicellular fossils are from the Ediacara Hills of Australia and date to just over 600 million years ago, near the close of the Precambrian Era. The Ediacara fossils include extinct species of invertebrates such as jellyfish, segmented worms, and sea pens.

At the beginning of the Cambrian Period of the Paleozoic Era invertebrate fossils become so abundant that this period is called the "Age of Invertebrates." Invertebrates are multicellular animals lacking an internal skeleton or vertebral column. Typical invertebrates of the Cambrian seas were the jellyfish, a great variety of shellfish, worms, and trilobites (a distant relative of the insects). Some typical invertebrates of the Cambrian ocean are pictured in Figure 8.5.

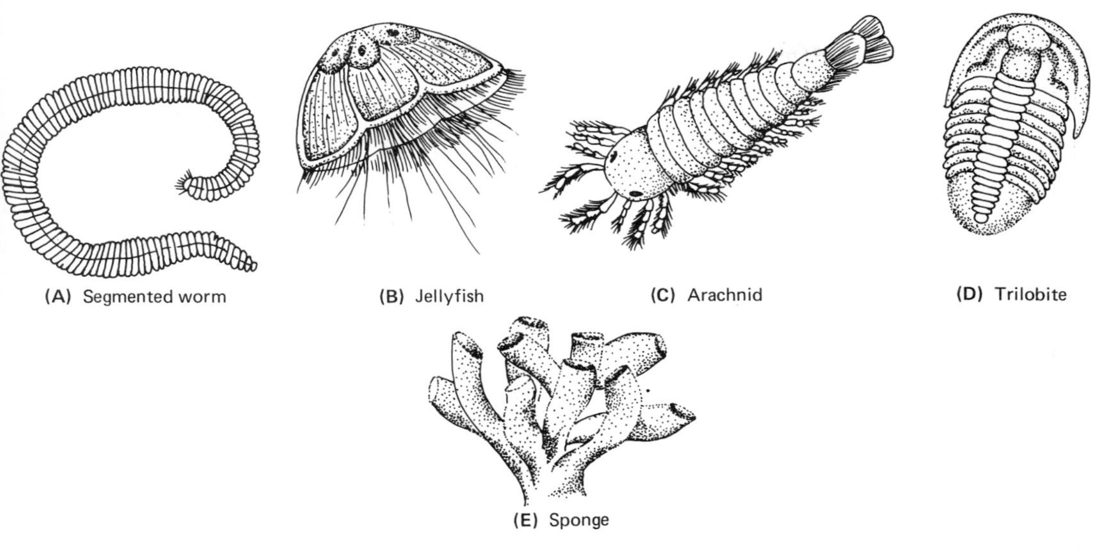


**Figure 8.5** Typical Cambrian invertebrates.

# 9 Phylogenetic Divergence

## Introduction

The Paleozoic Era witnessed the rise of the invertebrates and the origin and proliferation of three major groups of vertebrates: fish, amphibians, and reptiles. Two other vertebrate forms, the birds and mammals, arose in the Mesozoic. Each of these groups evolved distinct strategies for survival that permitted them to undergo adaptive radiation into a position of dominance over their rivals.

## Invertebrates

The first invertebrates appeared late in the Precambrian Era and, by the early stages of the Paleozoic, were the dominant animal forms in the waters of the world. The sudden appearance of large numbers of invertebrate forms presents an interpretive problem to paleontologists because the highly specialized body structures of these animals implies a much longer evolutionary history than is indicated by the fossil record. Two possible explanations for this situation have been offered: (1) invertebrates have a lengthy evolutionary history but their fossil record is incomplete, or (2) invertebrates actually underwent an explosive adaptive radiation near the end of the Precambrian Era.

The lack of older invertebrate fossils can be explained by the erosion of fossil-bearing strata or by assuming that early invertebrates lacked the hard body parts most commonly fossilized. These arguments are weakened by the fact that soft-bodied algae are found fossilized in older Precambrian rocks and that many of these algae lived in habitats similar to those occupied by invertebrates. It is difficult to explain why invertebrates, if they existed, were not interred along with these algae.

The other possibility is that invertebrates evolved from simple eucaryotic precursors late in the Precambrian and underwent a very rapid adaptive radiation because they had no natural enemies and had no strong competition for resources. A changing world climate during this period may also have been advantageous to the new life forms.

Whatever the explanation, the sudden appearance of a great variety of invertebrate fossils in early Paleozoic strata, along with a surprising lack of transitional forms, makes reconstruction of the evolutionary history of this group very difficult. Two possible patterns of invertebrate divergence are presented in Figure 9.1 but there is no strong evidence favoring either model.

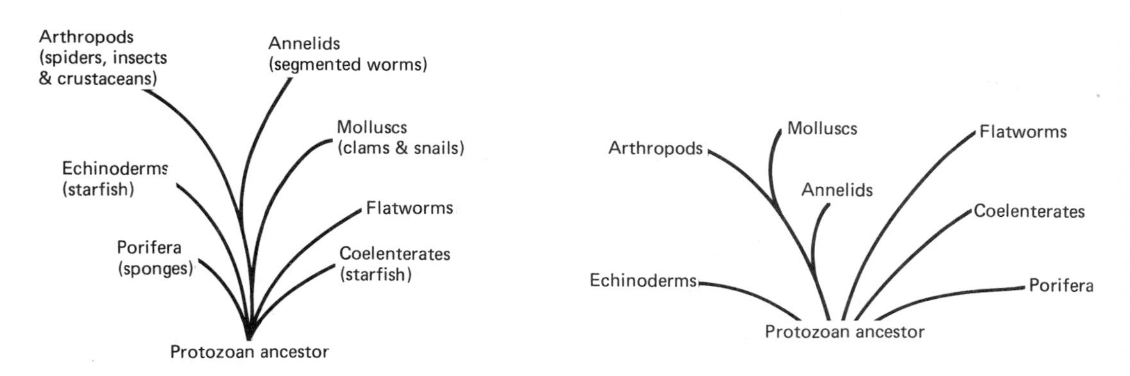


**Figure 9.1** Proposed patterns of invertebrate evolution.

# Vertebrates

Paleontologists generally agree that vertebrates evolved from the starfish line of invertebrates during the Cambrian Period because of the fishlike nature of the starfish larva and because of similarities between the protein chemistry of the starfish and vertebrates. It is thought that the earliest

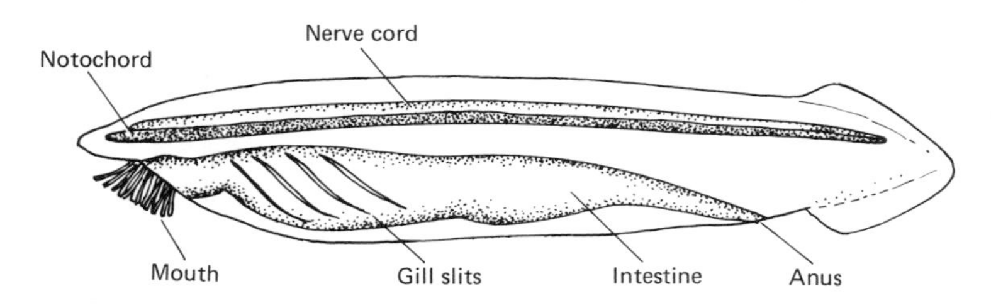


**Figure 9.2** The sea lancelet (*Branchiostoma*).

vertebrates resembled the sea lancelet (*Branchiostoma*), a small animal that presently inhabits coastal regions around the world (Figure 9.2). The lancelet is an anatomically simple animal that lacks a brain and has only a light-sensitive "eyespot" as sensory apparatus. But its fishlike body has a digestive tract that includes mouth, intestine, and anus, and it has gills that open into the pharyngeal ("throat") portion of the digestive tract as they do in fish. Most importantly, the lancelet has a *notochord* and a dorsal nerve cord. The notochord is a rodlike structure that runs the length of the body and provides both strength and support. It is present in some stage of the life cycle of all vertebrate animals.

## Early Fish

The first true vertebrate fossils appear in rocks of the Ordovician Period. These were the jawless fish (ostracoderms) that had a true spinal column enclosing the nerve cord and an internal skeleton of soft *cartilage* rather than bone (Figure 9.3A). Because they lacked jaws, the feeding habits of these fish were limited to sifting bottom sediments for organic matter or seining surface waters for floating aquatic organisms. Jawless fish lacked paired lateral fins and so were weak swimmers.

By the Silurian Period the jawless fish were joined by a more advanced group, the placoderms (Figure 9.3B). In these fish the bones of the third gill arch were modified to form a true jaw, which allowed for a wider range of feeding habits, including predation. The placoderms retained the cartilaginous skeleton of the jawless fish, and, as they had only imperfectly formed lateral fins, they were also weak swimmers.

Both the jawless fish and the placoderms were "armored fish"; the front portion of their bodies was covered with heavy plates that protected them against the large carnivorous invertebrates and predacious species of placoderms. Both groups were successful for a time, but their numbers dwindled during the Devonian Period and by the end of the Paleozoic Era the placoderms had become extinct. Only two lines of the jawless fish have survived into modern times—the lamprey eel and the hagfish. A major factor in the decline of the "armored fish" was the rise of the bony fish.

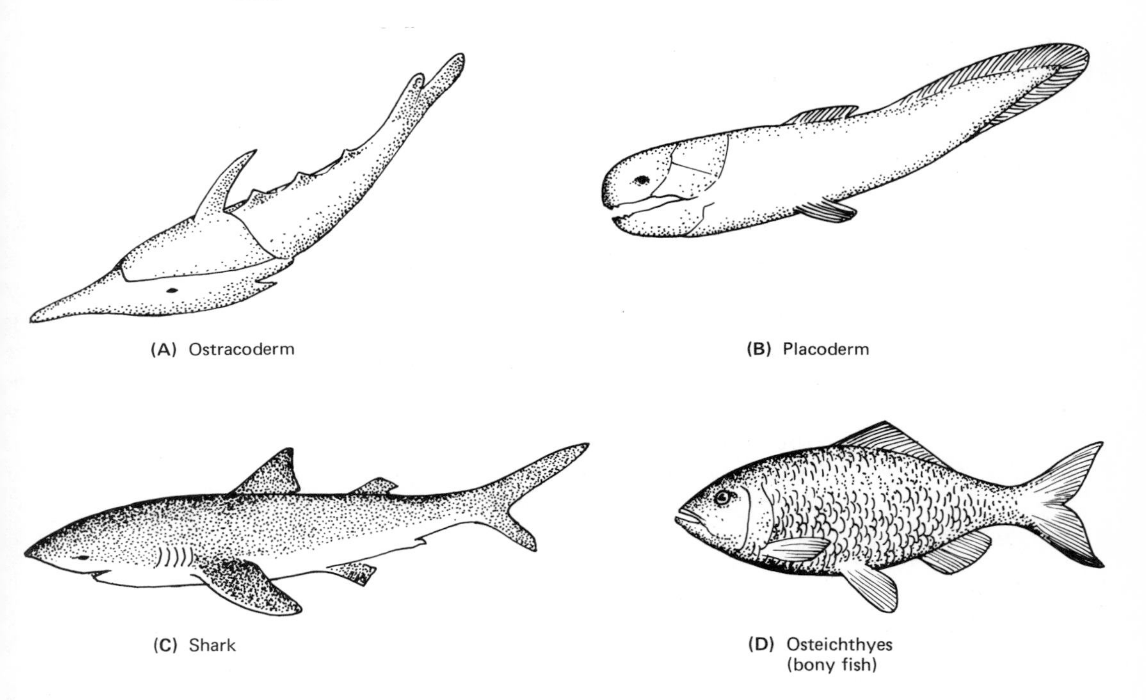

(A) Ostracoderm

(B) Placoderm

(C) Shark

(D) Osteichthyes (bony fish)

## Bony Fish

The bony fish (osteichthyes) were supremely well adapted to the aquatic environment, from their streamlined body shape to the presence of paired lateral fins that stabilized the fish in the water and provided for greater maneuverability (Figure 9.3C). The internal skeleton was strengthened by calcium deposits and the external plates of the armored fish were reduced to small, light-weight scales. The jaws were well developed, as were the teeth. The lightness, quickness, and strength of these fish made them more efficient swimmers and feeders than the slower and more cumbersome armored fish, and they quickly rose to a position of dominance over all other aquatic forms. The bony fish began an adaptive radiation in the Devonian Period that continues today.

By the end of the Devonian Period the bony fish had firmly established the success of the vertebrate line. From the innovation of the notochord had evolved a body plan based on an internal skeleton with a spinal column, and the calcification of this skeleton added great strength to the spine, skull, girdles, and fins—an important prerequisite for the transition to land.

## Sharks

Sharks (chondrichthyes) evolved at about the same time as the bony fish—first appearing in the Devonian and enjoying a period of expansion during the Carboniferous and Permian Periods (Figure 9.3D). Many forms became extinct toward the end of the Paleozoic and the surviving lines can be divided into two groups: (1) those that have the familiar streamlined, fish-shaped body, and (2) the skates and rays, which have flattened bodies and ventrally located mouths. All sharks are predatory and the group has remained almost exclusively marine in its distribution. The sharks are generally considered primitive fish because they have a cartilaginous rather than a bony skeleton.

## Transition to Land

Extensive droughts during the Devonian Period caused many streams and lakes to dry up and the sea level to fall. Many plants and animals were left stranded on drying mud flats, and although most perished, a few of these organisms possessed traits that allowed short-term survival out of water. These plant and animal species gradually accumulated traits that permitted them to survive for ever longer periods out of water, and they slowly became adapted to life on land.

The immediate problems faced by a fish out of water are: (1) how to obtain oxygen, (2) how to prevent water loss from its body surface, and (3) how to support itself and move. Fish breathe by means of gills, which are organs specialized to remove dissolved oxygen from water. Gills cannot function in air and a fish out of water literally suffocates. Fish also lack a protective covering to prevent water loss and soon dehydrate when exposed to the relative dryness of the atmosphere. Finally, the fins of the fish are neither properly structured nor positioned to provide for support and locomotion on land, and a fish left stranded by retreating waters cannot effectively move to find water.

## Lobe Fins and Amphibians

Among the many species of fish that populated the Devonian seas occurred the lobe fins (crossopterygians), a type of bony fish with an unusual combination of structures that favored its survival along the drying margins of the oceans.

The successful transition from water to land was made by the lobe fins because they had two important characteristics: (1) the unique structure of their lateral fins, and (2) the presence of lungs.

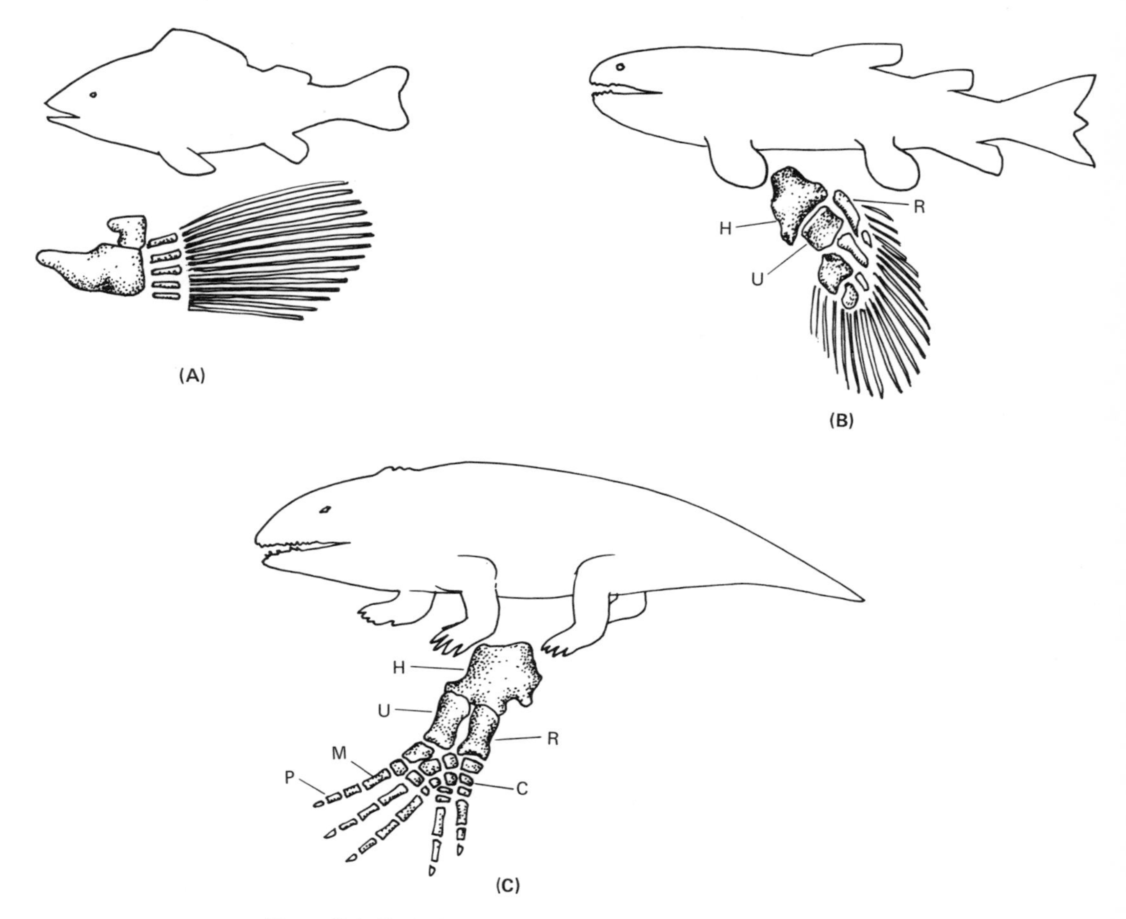


**Figure 9.4** Evolution of the amphibian forelimb. (A) Bone structure of perch fin. (B) Bone structure of lobe fin. (C) Bone structure of forelimb of the amphibian *Seymouria*.

In most bony fish the bones of the lateral fins radiated individually from the girdle, forming a fanlike structure well suited to their rudderlike function. In the lobe fins, a single bone extended from the girdle, followed by two shorter bones that lie side by side (Figure 9.4). This bone structure provided greater strength than the fin structure of other bony fish and permitted the lobe fin to "walk" for short distances—enabling them to move from drying areas to deeper and more permanent pools. The close resemblance between the bone structure of the lobe fin and the forelimb of *Seymouria*, one of the early amphibians, can be seen in Figure 9.4.

Some species of contemporary tropical fish possess air sacs and at times rise to the surface of the water to "gulp" air into these simple lungs. This behavior is especially prevalent when the water is very warm and

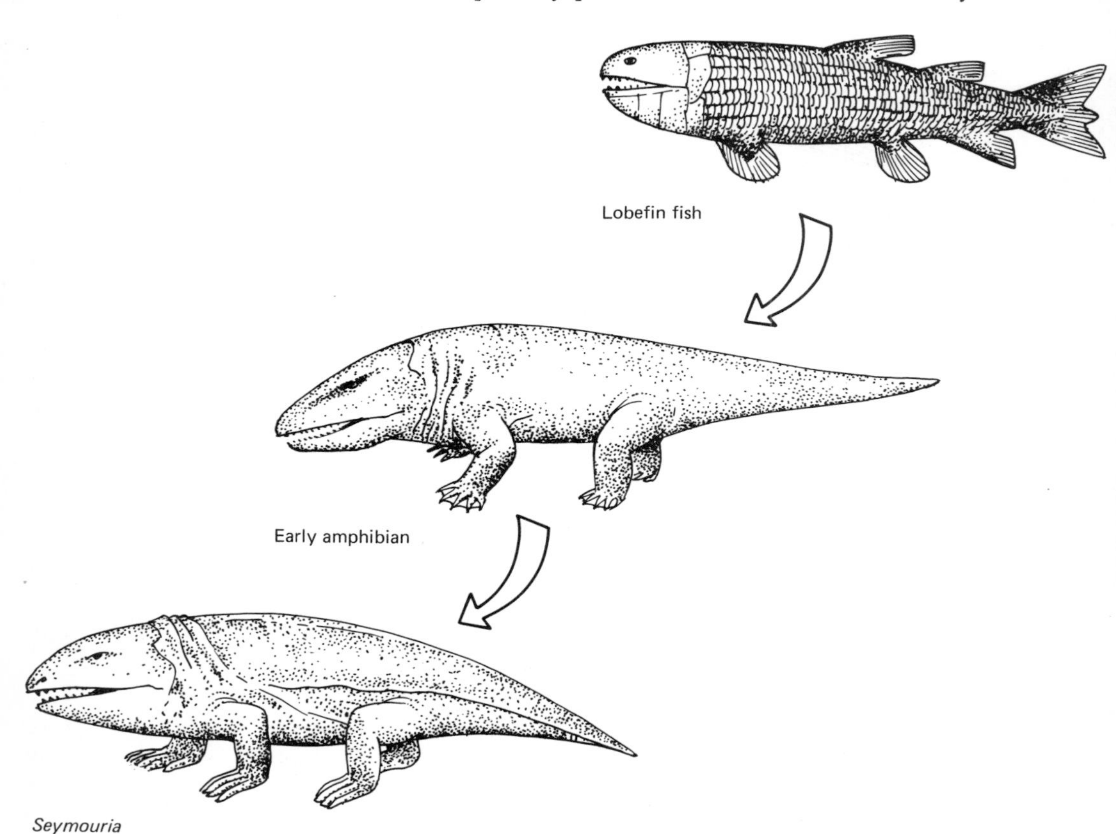


**Figure 9.5** Stages in the evolution of the amphibians.

dissolved oxygen levels are low. The ancient lobe fins had comparable structures, and probably used them in the same way. The ability to store air in primitive lungs would have aided in their survival out of water, and gradual modification of these structures into specialized organs capable of air breathing would have greatly extended the length of time they could survive on land.

In addition to adaptations for movement and air breathing, the lobe fins accumulated other traits that adjusted them to existence on land: (1) thicker skin to reduce water loss, (2) strengthening and modification of the spine, girdles, and limbs, (3) changes in the skull and teeth for feeding, and (4) improvement of the sense of smell and hearing, both more important in the terrestrial environment. The cumulative effect of these changes was the origin of the first group of terrestrial vertebrates, the amphibians (Figure 9.5).

The amphibians were very successful during the Carboniferous Period, a time when the continental plates were low and rainfall was plentiful, conditions that produced abundant food and shelter in the marshes and swamps that covered much of the land surface. Without danger from predators and without competition for resources the amphibians quickly spread over the land. But the amphibians were not completely terrestrial animals—they retained the external reproductive system of their fish ancestors and had to return to the water for mating. In external reproduction, the sperm and egg are released into the water and fertilization takes place externally. The developing zygote is surrounded by a thin membrane and nourished by a yolk rich in stored foods. This method of reproduction cannot function on land because of the extreme sensitivity of the gametes and zygote to drying.

## Reptiles

The generally warm and wet climate of the Carboniferous Period ended with a change to cooler and dryer weather—a climatic trend that, accompanied by elevation of the land masses, caused the drainage of wetland areas and the creation of dry upland habitats. With a reduction of favorable habitat, the water-oriented amphibians began to decline in number, a process accelerated by the rise of a new vertebrate group, the reptiles.

The reptiles were the first completely terrestrial vertebrates; they evolved from amphibian stock during the later Carboniferous Period and rose to dominance during the Permian Period. Physically, reptiles showed several advances over the amphibian line: (1) a deeper, narrower skull, (2) a tougher skin, (3) stronger spine and girdles, (4) slimmer legs, and (5) a

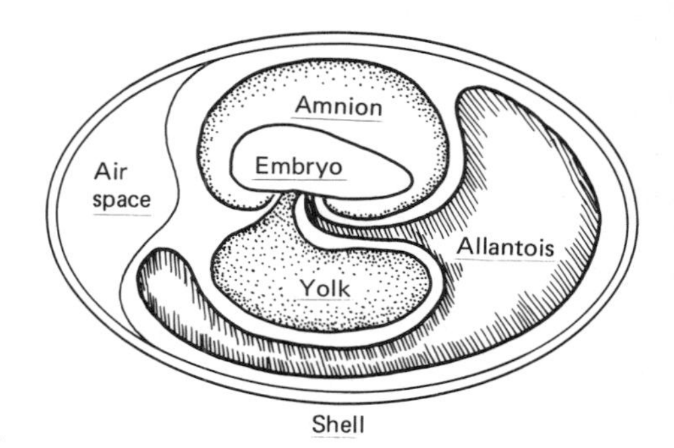


**Figure 9.6** The amniote egg.

larger brain. But, the most important advance of the reptilian line was the development of the *amniote egg.*

Reptiles evolved an internal reproductive system in which the sperm is introduced into the body of the female and fertilization takes place in the moist environment of the female reproductive tract. After fertilization, the zygote is surrounded by a membranous "water-sac" called the *amnion,* which maintains a moist environment for the developing embryo. Around the amnion is a second sac, the *allantois,* which receives and stores the waste products of metabolism. This entire structure is enclosed in a limey shell that permits the exchange of oxygen and carbon dioxide but prevents water loss (Figure 9.6). Because of this resistant outer shell, the amniote egg can be laid on land.

No longer required to return to water for reproduction, the reptiles colonized dry upland regions and invaded the swamps and marshes occupied by the amphibians. The first reptiles were the primitive cotylosaurs, the stock from which all other reptilian forms evolved. From this stock diverged a bewildering array of reptilian forms that occupied every conceivable terrestrial habitat, with two forms becoming aerial, while others returned to the sea as marine carnivores (Figure 9.7).

The successful adaptive radiation of the reptilian group reached its climax in the dinosaurs, a diverse group that produced the largest land creatures ever known to exist. While some dinosaurs were smaller than the modern crocodile, there was a definite tendency toward gigantism—*Tyrannosaurus rex,* the largest land carnivore that ever lived, was over 40 feet long, 20 feet tall, and weighed as much as 8 tons. Big as he was, *Tyrannosaurus* was dwarfed by the herbivorous *Brontosaurus,* which stood 80 feet tall and weighed 30 to 40 tons. Examples of the great variety of dinosaurs are presented in Figure 9.8.

The close of the Mesozoic Era marked the rapid extinction of reptilian forms. This great extinction is often attributed to another worldwide

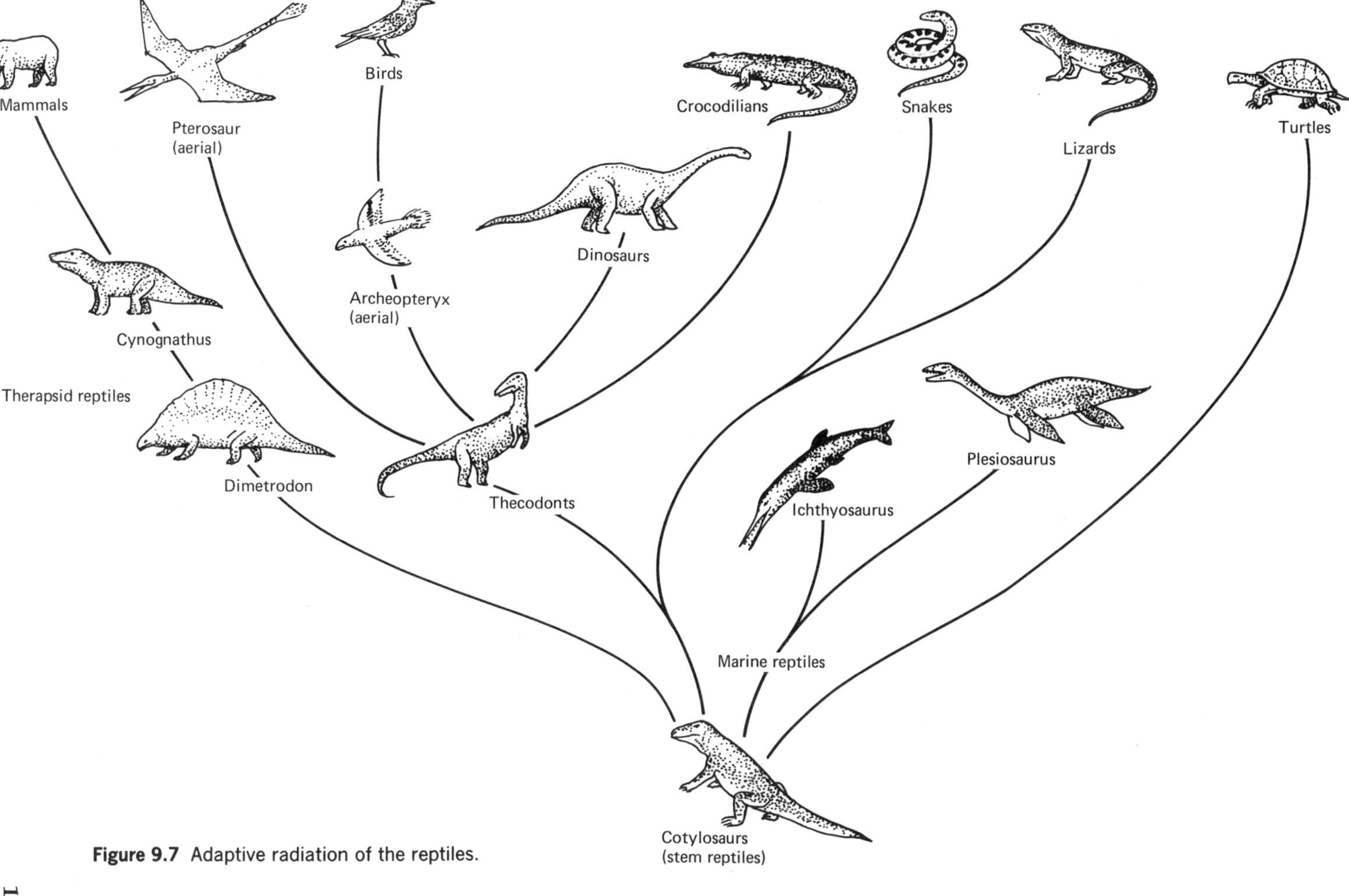


**Figure 9.7** Adaptive radiation of the reptiles.

**Figure 9.8** Dinosaurs. (A) *Brontosaurus;* (B) *Triceratops;* (C) *Tyrannosaurus;* (D) *Stegosaurus.*

climatic change, but some paleontologists feel the degree of change was too small to account for the disappearance of so many species. Other explanations are that a great influx of solar radiation was responsible for the extinctions, or that the dinosaurs were doomed by a growth trend in which their bodies became very large in proportion to their still-small brains—that is, the dinosaurs were no longer intelligent enough to compete and survive. Whatever the reason for the demise of so many species of reptiles, the only members of this once-great group of vertebrates to survive into recent times are the turtles, snakes, lizards, and crocodilians.

## Birds

Two groups of flying reptiles evolved during the Jurassic Period: (1) the pterosaurs (Figure 9.7), which remained a truly reptilian form and became extinct near the end of the Cretaceous Period, and (2) *Archeopteryx* (Figure 9.7), which evolved into a new vertebrate class, the birds. *Archeopteryx* is, in fact, classified with the birds because it possessed wings with feathers. Adaptations to flight in the birds include: (1) feathers replacing scales, (2) light, hollow bones, (3) powerful flight muscles, (4) keen eyesight, (5) a highly developed nervous system, and (6) a high metabolic rate to provide the energy needed for flight. The combination of insulating feathers and high metabolic rate contributed to the development of warmbloodedness, i.e., the ability to maintain a certain body temperature despite fluctuations in the ambient environment.

Both the ability to fly and warmbloodedness are *generalizing adaptations,* traits that permit an animal to survive in a wide range of habitats. Flight allowed birds to escape threatening conditions and warmbloodedness greatly expanded the range of environmental conditions in which the birds could live. Coldblooded reptiles and amphibians could survive cold weather only by hibernation or a limited migration, and thus cold weather closely circumscribed their activities. While some reptile groups may have developed warmbloodedness this trait did not survive in the reptilian line beyond the Mesozoic.

The birds have been very successful in their adaptation to the aerial habitat and have spread over the continents and waters of the world. While modern birds display a great variety of size, shape, and life-style, there is a high degree of structural similarity among all birds, which dates to early Cenozoic times.

## Mammals

The first undoubted mammal fossils appear in rocks of Jurassic age, descendants of the *Therapsid* line of reptiles (Figure 9.7). Early mammals were small creatures, sharing the land with the ruling reptiles and remaining a minor part of the world fauna until the early Cenozoic, when they began an adaptive radiation that has made them the dominant vertebrate form of the land surfaces of the recent geologic period. Adaptive advances of the mammals include: (1) hair covering the body, (2) internal development of the embryo, (3) mammary glands for feeding the young, (4) warmbloodedness, and (5) a tendency toward larger brain size and increased intelligence. As in the birds, warmbloodedness allowed the mammals to extend their geographic range and remain active the year around. Specific adaptations to warmbloodedness in mammals are the insulating effect of hair and an increased metabolic rate.

In all mammals except the monotremes, the embryo is retained within the body of the female parent during at least the early stages of development. All mammals feed their young from specialized mammary glands. These advances in reproduction have had an important effect on mammalian behavior because they necessitated a prolonged relationship

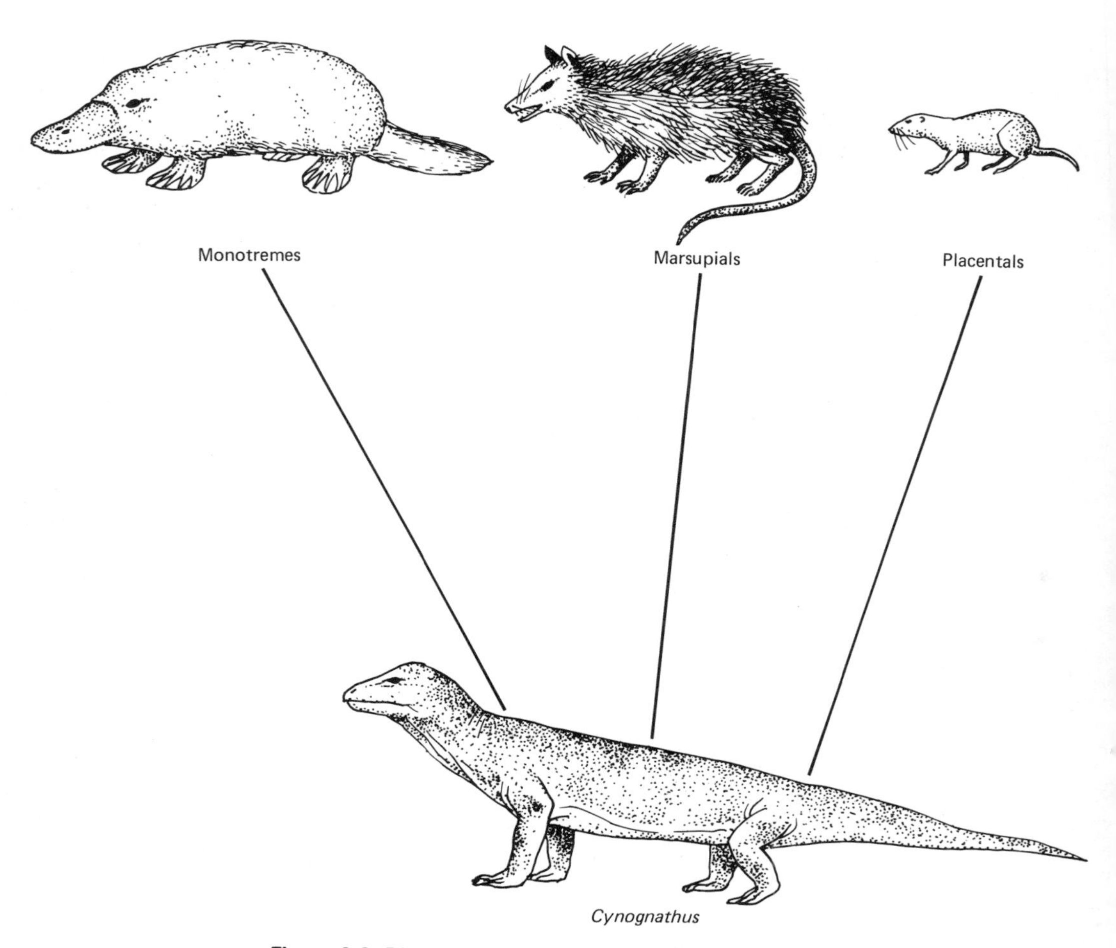


**Figure 9.9** Divergence of mammalian forms. *Cynognathus* is representative of the therapsid reptiles from which the first mammals evolved. The monotremes are primitive mammals that first appeared in the Jurassic Period; the marsupials and placentals appeared in the Cretaceous Period.

between the mother and her infant. The origin of family and group behavior are traced, in part, to these reproductive advances.

Another major advance of the mammalian line is the increase in brain size in proportion to body size and the increase in intelligence that has accompanied it. The success of the mammals is ascribed as much to superior intelligence as to any of their physical attributes.

Of the groups of mammals that evolved during the Jurassic and Cretaceous Periods, three lines have survived into recent times: (1) monotremes, (2) marsupials, and (3) placentals (Figure 9.9).

## Monotremes

Monotremes are found only in Australia and New Guinea where they are represented by three genera—the duckbill platypus (*Ornithorhyncus*) and the spiny anteaters (*Echidna* and *Tachyglossus*) (Figure 9.10). Monotremes are primitive animals, having retained elements of reptilian skeletal anatomy and the reptilian habit of laying eggs. But, they are warmblooded, covered with hair, and have primitive mammary glands for feeding their young. The monotremes of today are little changed from ancestral forms that evolved from reptilian stock in the Jurassic.

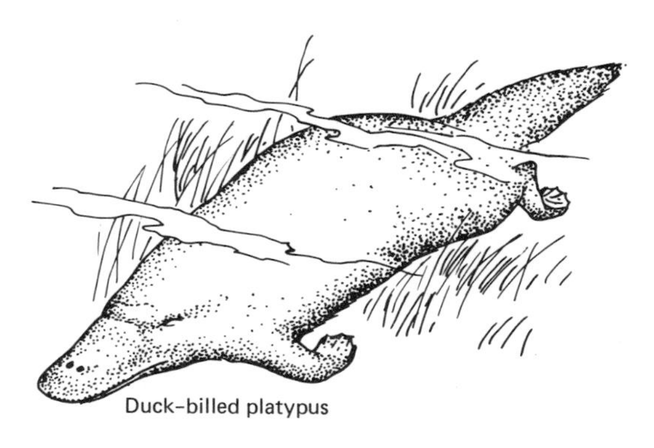


**Figure 9.10** Monotremes.

## Marsupials

Marsupials (Figure 9.11) are more "mammal-like" than monotremes in that: (1) the skeleton is more mammalian, (2) the mammary glands are

**Figure 9.11** Marsupial mammals.

discrete, specialized structures, and (3) the marsupial embryo develops internally. However, the skull of the marsupial is not large, reflecting only moderate brain development, and the young are born at an early stage of development and must migrate to the marsupium (pouch) of the mother where further development takes place. Within the marsupium, the young are fed from mammary glands.

Marsupial mammals preceded placental mammals in undergoing adaptive radiation and were widely spread over Laurasia and Gondwanaland, the two land masses existing in the Cretaceous Period (Figure 9.12). When these land masses separated into individual tectonic plates many

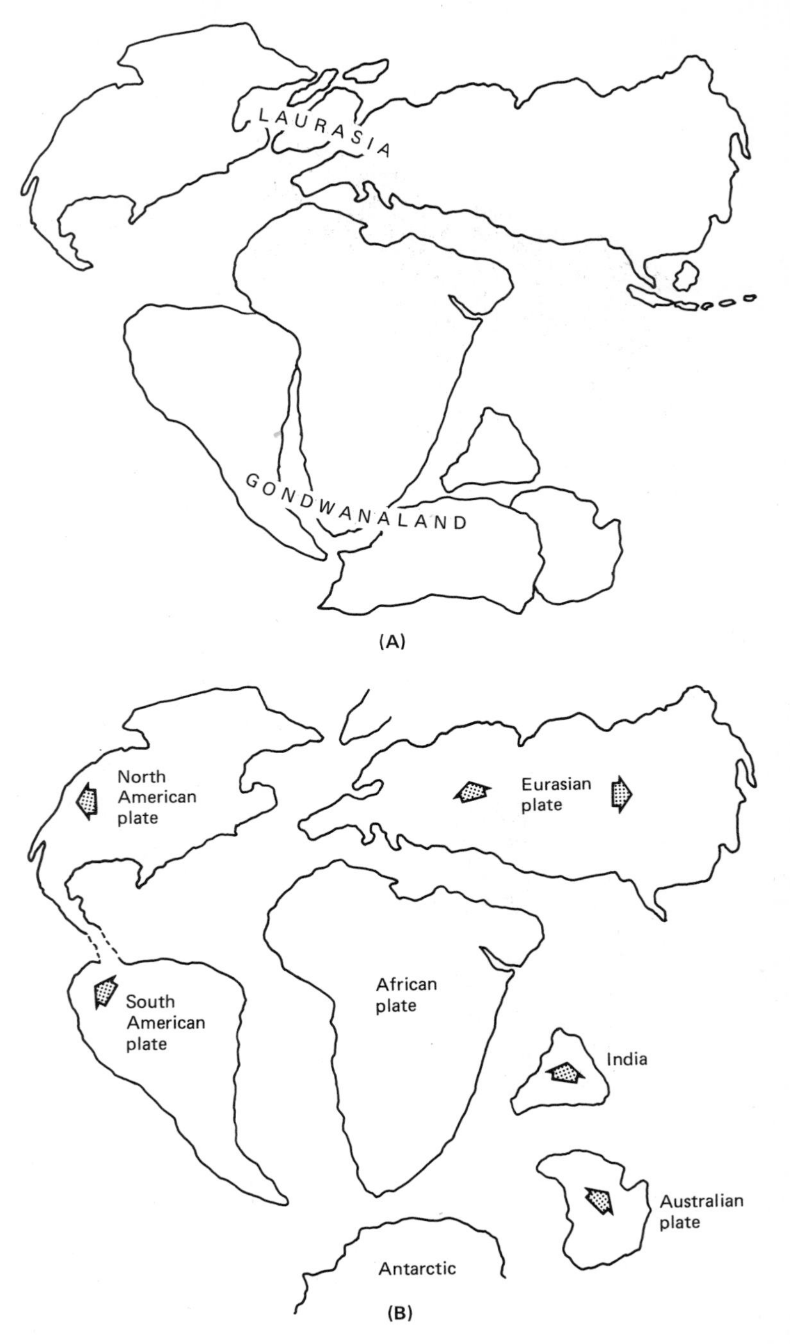


**Figure 9.12** Land masses of the Cretaceous Period. At the beginning of the Cretaceous Period, the continental plates were united in two major land masses: Laurasia and Gondwanaland (A). During the Cretaceous Period, these land masses broke apart and the individual plates began their migrations to their present positions (B). Arrows indicate direction of motion.

**Figure 9.13** Major groups of placental mammals. (A) *Perissodactyla*, including zebras, horses, rhinoceroses, and tapirs; (B) *Chiroptera*, including the bats; (C) *Proboscidea*, including the elephants; (D) *Primates*, including lemurs, lorises, monkeys, apes, and humans; (E) *Lagomorpha*, including rabbits and hares; (F) *Edentata*, including armadillos, anteaters, and sloths; (G) *Carnivora*, including bears, wolves, cats, otters, and skunks; (H) *Insectivora*, including the moles and shrews; (I) *Cetacea*, including the whales and dolphins; (J) *Artiodactyla*, including giraffes, reindeer, pigs, camels, hippopotamuses, and deer; and (K) *Rodentia*, including rats, mice, squirrels, beavers, and porcupines.

(G)
(F)
(H)
(I)
(K)
(J)

mammalian populations became geographically isolated, initiating a period of divergence and speciation that played a major role in the future evolution and distribution of mammalian types. The placental mammals were just beginning their expansion at this time and none had reached the Australian region before it separated and began its slow drift northward. In this isolated region marsupials and monotremes developed in protective isolation from the placentals.

Marsupials also remained the dominant mammalian form on the South American continent through the Tertiary Period because this tectonic plate also separated before placental mammals entered the region. However, in the Paleocene Epoch the Central American land bridge was established between the North and South American continents and placental mammals invaded from the north and exterminated many marsupial species. Today, only a few species of marsupials survive in the Americas—the only North American species is the opossum (*Didelphis virginiana*).

## Placentals

In general, mammalian traits have reached their highest development in the placentals. They have a larger brain size and are considered the most intelligent mammalian group. The placental embryo is retained within the mother for a longer period than in other mammals. Food and wastes are exchanged through the umbilical cord which is attached to the uterus by the placenta. In all regions of the world, except Australia, placentals are the dominant mammalian form.

There are 16 orders of placental mammals ranging from marine carnivores (seals) to flying insect-eaters (bats). The major groups of placental mammals are pictured in Figure 9.13.

## Plant Evolution

Plants are of fundamental importance to the perpetuation and expansion of animal life. The green plant is the ultimate food source for all animals, and the oxygen released by the process of photosynthesis is essential to life. Beyond this, plants provide shelter and nesting sites for animals, aid in the geologic process of soil formation, and help stabilize the soil once it has been formed. The transition of animals to land could not have been successful without the successful establishment of plants on the land surface.

The plant kingdom can be divided into two broad groups: (1) nonvascular plants, and (2) vascular plants. The presence or absence of vascular tissue is an important distinction because the development of an internal conductive system was one of the major factors in the colonization of the terrestrial habitat. The vascular system is composed of two types of tissue: *xylem* and *phloem.* Xylem tissue transports water and dissolved minerals throughout the plant, and phloem tissue transports food made in the leaves to nonphotosynthetic parts of the plant body. In addition to their function in water transport, xylem cells have thickened walls that provide strength to hold the plant body erect. The hard, woody part of trees is xylem tissue (Figure 9.14).

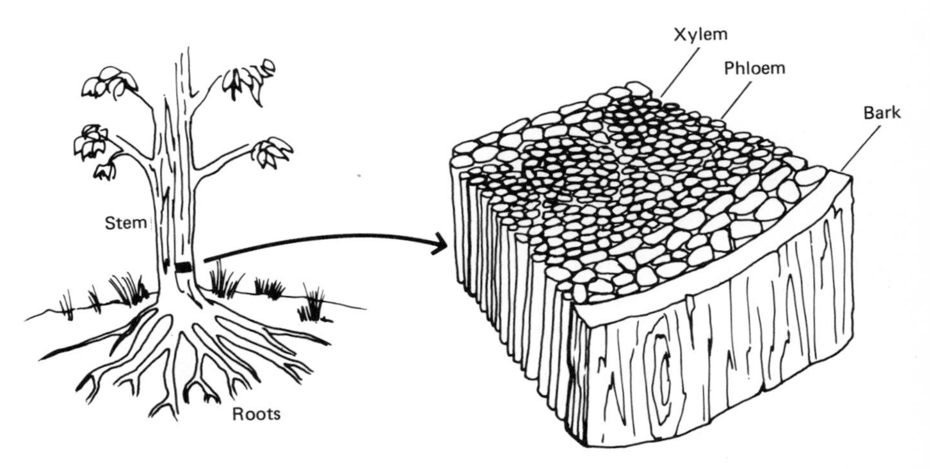


**Figure 9.14** Adaptations of vascular plants to the terrestrial environment. Cross-section of a portion of stem showing xylem and phloem.

## Algae

The first plants were procaryotic blue-green algae that date to over three billion years ago. Eucaryotic green algae appeared later, about one and one-half billion years ago (see Chapter 8). Other algal forms include the red algae, brown algae, and diatoms. The five groups differ in their pigments (and thus color); the chemistry of their cell walls; and type of food storage molecule—which ranges from oils to different types of carbohydrate molecules.

All algae are photosynthetic and nearly all are aquatic—although a few live in moist terrestrial habitats. Algae are nonvascular plants that lack differentiation into roots, stems, and leaves. Their level of organization ranges from single-celled to colonial to filamentous. Filaments are strings of cells joined end-to-end and individual filaments may be either branched or unbranched. There is little cellular specialization in the algae and sex organs are simple and are not usually surrounded by sterile cells (Figure 9.15).

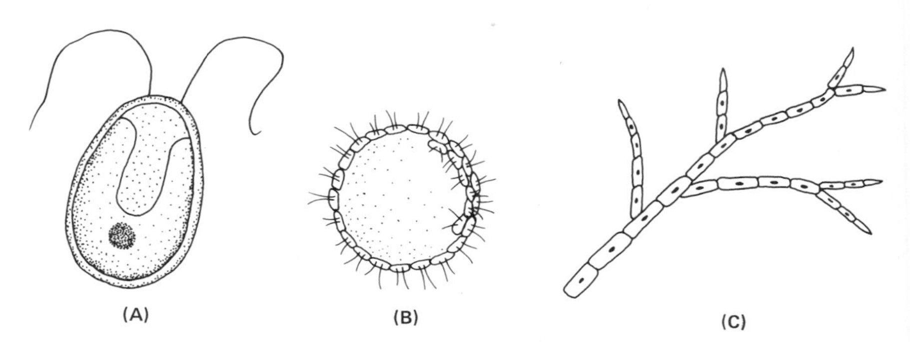


**Figure 9.15** Levels of organization in the algae. (A) The single-celled alga, *Chlamydomonas,* (B) The colonial alga, *Volvox,* and (C) The filamentous alga, *Stigeoclonium.*

## Life Cycles

Three kinds of life cycles occur among the algae and it is important that these are understood before we discuss other plant groups or the events of plant evolution. One life cycle is similar to that presented in Chapter 4 (see Figure 4.1): the free-living stage of the life cycle is spent in the diploid condition and haploid gametes are formed by meiosis (Figure 9.16A). Fertilization results in a diploid zygote that grows into a new adult organism. This resembles the life cycle of animals.

A second life cycle present in some algae varies from the one described above in that the adult, free-living stage of the life cycle is haploid rather than diploid (Figure 9.16B). In this life cycle the haploid plant produces gametes by mitosis. After fertilization, the diploid zygote divides meiotically to produce haploid cells—each of which grows into a new, free-living haploid plant.

The third life cycle is the most complex. It is generally called *alternation of generations* (Figure 9.16C). There actually are two distinct generations in the life cycle: (1) a diploid sporophyte generation, and (2) a haploid gametophyte generation. The sporophyte produces spores by meiosis. Upon germination these spores grow into haploid gametophytes. The gametophyte produces gametes by mitosis and the gametes fuse to form a zygote that grows into a new sporophyte. Alternation of generations is present in all land plants, both vascular and nonvascular.

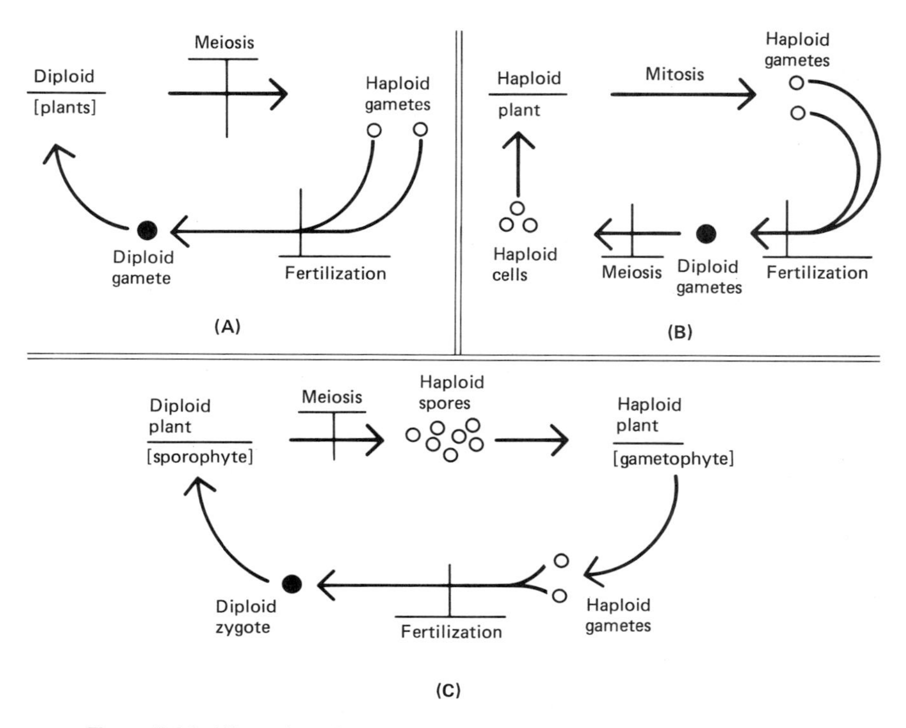


**Figure 9.16** Life cycles of the algae.

# Bryophytes

Bryophytes (mosses and liverworts) are structurally more complex than the algae but simpler than vascular plants. The bryophytes do not have an internal vascular system and have *rhizoids* rather than roots. They are generally low-growing plants that are limited in their distribution to moist habitats. Within these areas bryophytes are often quite prolific, forming extensive mats over the substrate. Moss gametophytes consist of a leafy axis that may grow either upright or prostrate, and which is anchored in the soil by the rhizoid (Figure 9.17A).

The life cycle of the moss displays alternation of generations with the gametophyte being the conspicuous, free-living phase (Figure 9.17B). The moss gametophyte bears two types of sex organs: (1) an antheridium, which produces the male gamete, and (2) an archegonium, which produces the egg. The male gamete is motile and requires a film of water to swim to the archegonium to accomplish fertilization of the nonmotile egg cell. The diploid zygote undergoes division within the archegonium and the sporophyte plant remains permanently attached to the gametophyte.

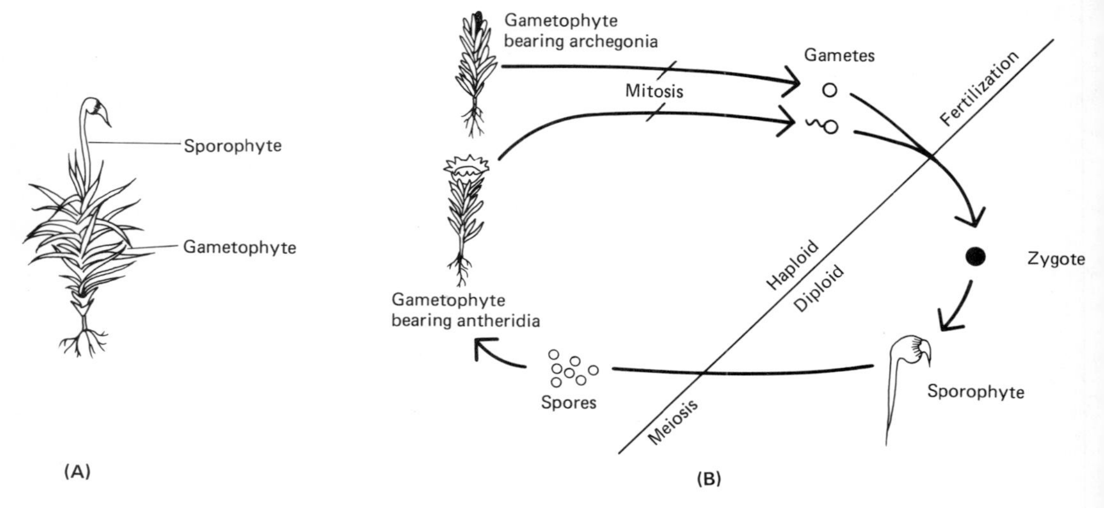


**Figure 9.17** Mosses. (A) Moss plant, with sporophyte growing out of gametophyte. (B) Life cycle.

The mature sporophyte produces haploid spores by meiosis. These spores fall to the ground where they initiate new gametophyte plants.

## Vascular Plants

Vascular plants first appeared in the Silurian Period and proliferated during the Devonian and Carboniferous Periods. All representatives of this line of descent have specialized conductive cells called *tracheids.* The first vascular plants were structurally simple, lacking roots and leaves. In these plants, called *psilopsids,* the stem served as the organ of photosynthesis and an underground stem functioned as a root (Figure 9.18A). Gradually, these early vascular plants accumulated traits that better adapted them to the dryness of the terrestrial environment. Flattened structures developed along the stem that evolved into leaves—organs specialized to carry on photosynthesis. The underground stem became differentiated to form a root—an organ adapted to stabilization of the plant in the soil and to the uptake of water and dissolved minerals. Another special adaptation to the dryness of the terrestrial environment was the development of a waxy cuticle that covers the aerial portion of all vascular plants and greatly reduces water loss through evaporation.

The psilopsids were joined in the Silurian Period by a group of vascular plants called lycopsids (club mosses)—a group that, despite its com-

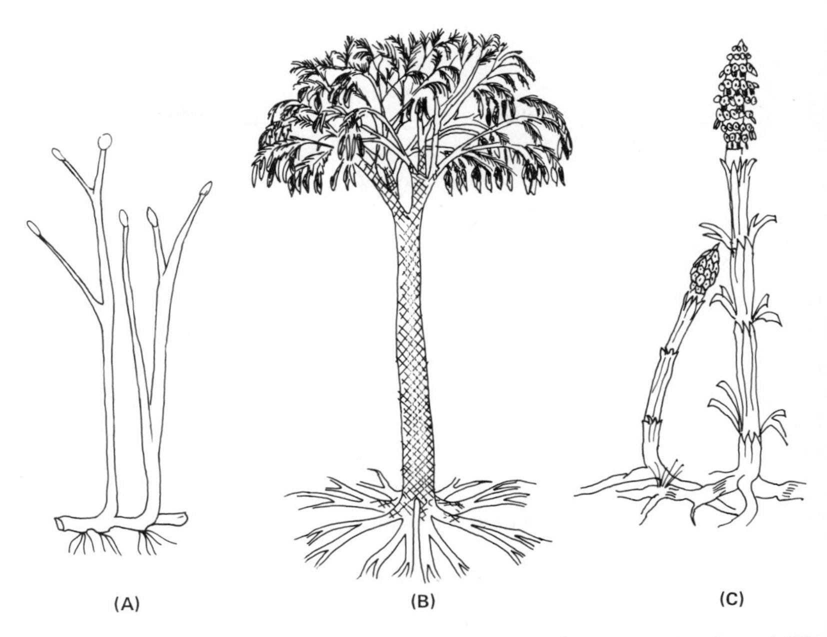


**Figure 9.18** Early vascular plants. (A) Psilopsid, (B) Club moss (lycopsid), and (C) Horsetail (sphenopsid).

mon name, is not related to the bryophytes (Figure 9.18B). Club mosses dominated the flora of the Devonian and Carboniferous Periods, with some members attaining tree size, forming the first forests on the Earth's surface. A third group of primitive vascular plants, the sphenopsids (horsetails), appeared in the Devonian Period (Figure 9.18C) and a fourth group, the pteropsids (ferns), in the Carboniferous Period.

## Pteropsids

Pteropsids (ferns) range in size from the small, delicate plants found in temperate climes to the big-leafed tree ferns of the tropics. Ferns have a well-developed vascular system; clear differentiation into roots, stems, and leaves; and a life cycle dominated by the diploid sporophyte. The fern sporophyte has specialized leaves, called *sporophylls,* which bear structures called *sporangia.* Within the sporangia, haploid spores are formed by meiosis. These spores fall to the ground where they germinate and grow into small, heart-shaped gametophytes—a free-living stage that, when mature, produces both antheridia and archegonia on a single gametophyte (in some ferns antheridia and archegonia are produced on sepa-

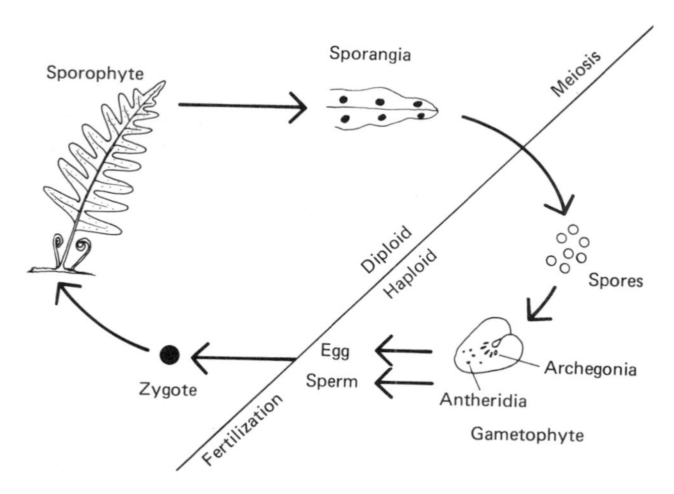


**Figure 9.19** Life cycle of a fern.

rate gametophytes). The nonmotile egg cell is retained within the archegonium and the motile sperm cell requires a film of water to swim to the egg and accomplish fertilization. The zygote gives rise to a new, free-living sporophyte plant (Figure 9.19).

The life cycle of the fern is typical of primitive vascular plants in that (1) there are disjunctive, free-living sporophyte and gametophyte generations; (2) the sporophyte generation dominates the life cycle; (3) spores are produced in sporangia borne on sporophylls; (4) each gametophyte produces both antheridia and archegonia, and (5) the sperm cell is motile and requires a film of water for successful completion of the fertilization process.

In advanced vascular plants—the pteridosperms, gymnosperms, and angiosperms—the evolutionary tendency is toward: (1) a reduction in the size and complexity of the gametophyte; (2) production of two types of gametophytes—one that produces a male gamete and one that produces a female gamete; (3) specialization of the sporophylls; (4) production of a pollen grain, and (5) development of the seed.

The evolution of the pollen grain and the seed are as important to plant evolution as the development of internal fertilization and the amniote egg was to vertebrate evolution. The pollen grain is a resistant structure that contains the male gamete and makes possible the accomplishment of fertilization without the presence of water. The seed contains an embryonic plant within a protective seed coat. Taken together, these adaptations permitted the successful migration of vascular seed plants from low marshy regions to dryer upland habitats.

# Pteridosperms

The first seed plants to appear in the fossil record were the pteridosperms—called "seed ferns" because of their superficial resemblance to the true ferns. However, the pteridosperms produced pollen and seeds and thus were evolutionarily more advanced than the ferns. The pteridosperms first appeared in the Carboniferous Period and became extinct in the Cretaceous Period.

# Gymnosperms

Gymnosperms (e.g., gingkos, conifers, and cycads) are evolutionarily advanced plants in which the life cycle is dominated by the sporophyte generation—the gametophytes being reduced to a few cells. The name gymnosperm literally means "naked seed," referring to the fact that the seeds of these plants are not enclosed in a fruit as they are in the angiosperms.

The conifers (pines, yews, larches, and their relatives) are shrub- and tree-sized woody plants with needle-shaped leaves. Because most conifers do not shed their leaves annually these plants are commonly called "evergreens." The sporophylls of the conifers are modified into two types of cones—one that produces pollen grains and one that produces ovules (Figure 9.20). The ovules are borne on the scales of the cone and each ovule is a megasporangium, inside which is a diploid megaspore mother

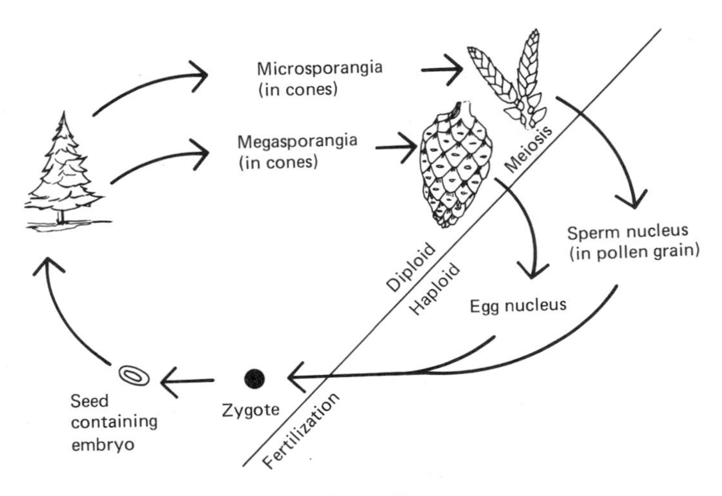


**Figure 9.20** Generalized life cycle of a conifer.

cell. The mother cell divides meiotically to produce four haploid daughter cells. Three of these cells disintegrate and the fourth undergoes several mitotic divisions, forming the female gametophyte. The female gametophyte bears two to five archegonia, each of which contains a single egg cell.

The second type of cone bears microsporangia under its scales. The microsporangium contains a microspore mother cell that divides meiotically to produce haploid microspores. Each microspore produces a pollen grain—a two-celled male gametophyte that is resistant to drying and that is carried by the wind to the ovule-bearing cone. One nucleus of the pollen grain controls development of a pollen tube, which elongates, enters the archegonium, and makes contact with the egg cell. The other (sperm) nucleus is then discharged into the egg cell and fuses with the egg nucleus (fertilization).

After fertilization, the zygote undergoes mitotic division, forming an embryo, which is surrounded by the tough, protective seed coat and is capable of surviving adverse environmental conditions. When favorable environmental conditions return the seed germinates, producing a new sporophyte. Many seeds contain stored food that is used by the germinating embryo until the growing seedling becomes photosynthetic.

Major advances in the reproductive cycle of the conifers include (1) modification of the sporophylls into cones; (2) reduction of the gametophyte generation; (3) production of separate male and female gametophytes; (4) incorporation of the male gametophyte in the pollen grain; and (5) formation of a seed.

## Angiosperms

The angiosperms (flowering plants) are the most advanced vascular plants. They generally have broad, flat leaves that, in temperate climates, they shed annually. The major reproductive advances of the angiosperms are the flower and the fruit. The flower contains petals, sepals, stamens, and pistils. Not every flower contains all of these structures—in some species the stamens and pistils occur on separate flowers. Pollen grains are produced in the stamens and the egg cell within the pistil. As in the conifers, the gametophyte generation is greatly reduced and separated into a female and a male component.

In some angiosperms the petals and sepals of the flower are bright-colored, and, along with the sweet nectar of the flower, attract insects and birds, which serve as agents of pollination—the pollen grains adhering to their bodies as they move from flower to flower. In other angiosperms, the flowers are less conspicuous and pollen is transferred from flower to

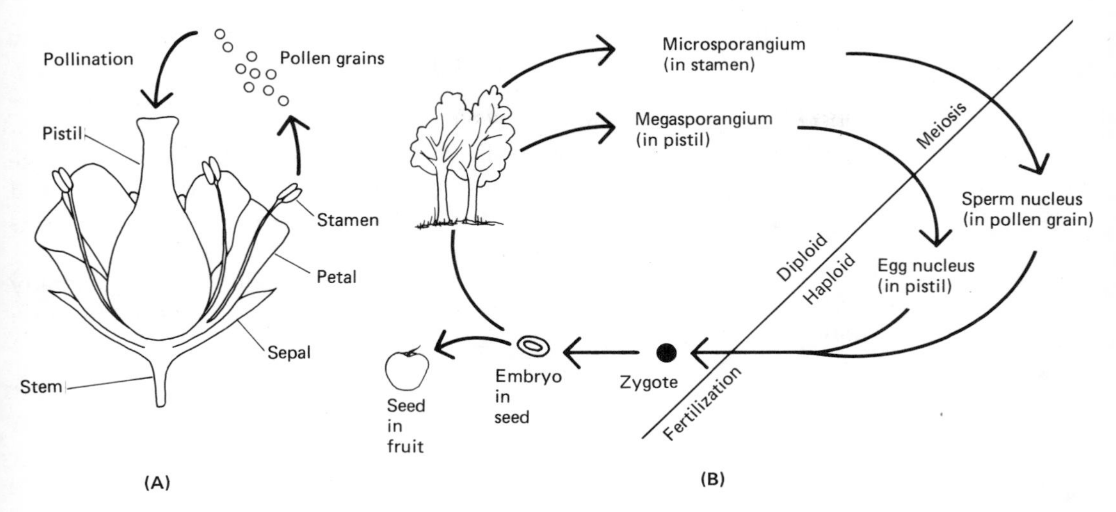


**Figure 9.21** Generalized life cycle of an angiosperm.

flower by the wind. Once the pollen grain reaches the upper surface of the pistil a pollen tube is produced and the sperm nucleus moves through the pollen tube and fertilizes the egg nucleus (Figure 9.21).

Upon fertilization, the zygote divides mitotically, forming an embryo, which is enclosed within a seed coat. The pistil of the flower enlarges and matures to form a fruit—which varies from the soft fleshy structure of the apple or peach to the dry hardness of a hickory nut. The fruit provides protection for the seed and also aids in seed dispersal—animals eat the fruit but often the seed passes through the digestive tract undisturbed and may be deposited with the animal's feces over a wide geographic range. The seeds themselves are important to other animals, providing a source of stored food that is available throughout the year. Seed crops, especially the grasses (e.g., corn, wheat, and rye), have played an important part in the biological and cultural history of the human race.

## Phylogeny

Many details of plant evolution remain uncertain because of a scanty fossil record and difficulty in interpretation of the fossil material available. The first plants were the procaryotic blue-green algae, but it is not certain whether the eucaryotic algal forms (greens, reds, browns, and diatoms) evolved from the blue-greens or arose polyphyletically. It is generally agreed that both the bryophytes and the vascular plants arose from the green algae. The bryophytes appeared in the Carboniferous Period and have given rise to no other plant form.

The vascular plants first appeared in the Silurian Period and their relationship with the green algae is based on similarities in (1) kinds of chlorophyll pigments, (2) cell-wall chemistry, (3) carbohydrate storage molecule (starch), and (4) the motile sperm cells of the green algae and primitive vascular plants. The first vascular plants were the psilopsids from which diverged the club mosses, horsetails, and ferns.

The ancestral stock of the seed plants is uncertain, but it is generally believed that the pteridosperms, gymnosperms, and angiosperms arose polyphyletically.

# 10 Human Evolution One

## Introduction

Humans are classified with the primates, a mammalian group that began its divergence during the Paleocene Epoch of the Cenozoic Era. Primates descended from mouse-sized terrestrial mammals that entered the trees either to escape predators or to seek the insects that made up their diet. These animals prospered in the trees, gradually accumulating adaptive structures that made them truly arboreal (tree dwellers): the bones of the limbs became elongate and slightly curved; the shoulder and hip joints became more flexible, allowing for greater freedom of movement; and the thumb and "great toe" were separated from the other four digits, permitting the animal to grip branches for security and movement—and to manipulate objects with its hands.

The sensory apparatus changed because the sense of smell, so important on the ground to detect predators and find food, was less important in the trees. On the other hand, an acute sense of sight and depth perception are essential to tree dwellers and, while the nose was reduced in size, the eyes became larger and moved to the front of the skull so that their fields of sight overlapped, producing stereoscopic vision.

Other trends in primate evolution have been the flattening of the face

**Figure 10.1** Trends in primate evolution. As primitive insectivores entered the trees they underwent specific adaptations to the arboreal habitat, including shortening of the muzzle; rounding of the skull; migration of the eyes to the front of the head; elongation of the limbs; development of prehensile hands, feet, and, in some cases, tails.

and rounding of the skull. The flattening of the face came about because the size and number of teeth were reduced, reflecting a dietary shift from insects toward the softer fruits and leaves found in the trees. As the teeth and nose were reduced, the muzzle became shorter and the plane of the face flatter (Figure 10.1). The rounding of the skull resulted from an increase in brain size. The brain centers controlling vision, the manipulative activities of the forelimbs, and intellectual development have undergone a distinctive increase in size and, as the brain grew larger, the cranial portion of the skull enlarged as well. In general, the primates are characterized by their activity, curiosity, and intelligence.

# Evolution of the Primates

The earliest primates probably resembled the oriental tree shrew (*Tupaia*) that lives today in the forests of southeast Asia. Two lines of descent evolved from primitive tupaioid stock; the prosimians and the anthropoids. Modern representatives of the prosimian line are the lemurs, tarsiers, and lorises of the forests of Asia and Africa (Figure 10.2). The three modern groups of anthropoids are: (1) New World monkeys, (2) Old World monkeys, and (3) the great apes and humans. New World monkeys are found in the forests of Central and South America and are characterized by their prehensile (grasping) tails. They are the most primitive group of living anthropoids and, because of their meager fossil record, their evolutionary relationship to other anthropoid lines is unclear.

The Old World monkeys are now confined to Africa and Asia but once occupied Europe as well. Their fossil record is more complete, dating to Oligocene strata of the Fayum district of Egypt. One of the Fayum fossils, *Parapithecus,* was a small, monkeylike creature thought to be ancestral to the Old World monkeys. The earliest ancestor of the line leading to modern apes and humans is also found in the Fayum strata, but before we discuss the history of this anthropoid group we should first examine the criteria used to distinquish man from his closest relatives.

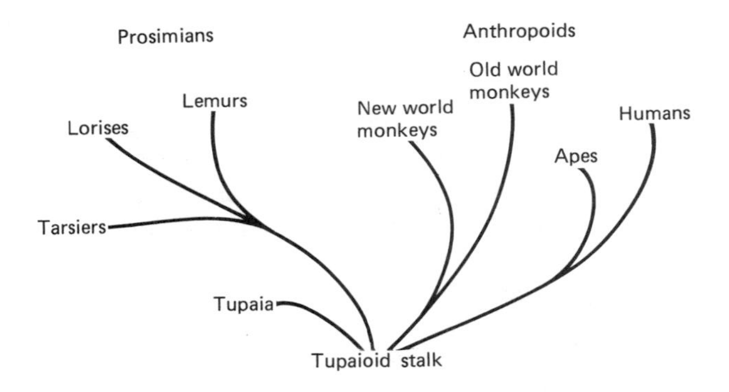


**Figure 10.2** Divergence of the primates.

# Being Human

A fundamental problem faced by scientists in their attempt to reconstruct the course of human evolution is to define just what is meant by "being human." What is there in a few fossil teeth, a fragment of skull, or a segment of hip bone that tells us whether the creature that possessed them in life should be classified as ape or human? To resolve this interpretive problem the paleontologist and anthropologist must consider a variety of physical traits and weigh them against modern representatives of these two groups. For convenience we can divide these physical attributes into two broad categories: (1) those associated with the skull, and (2) those associated with the post-cranial skeleton—the spine, girdles, and limbs. Occasionally the scientist is aided in this difficult task by finding some evidence of cultural advance such as tools, a hearth, or art work which indicate the advanced intelligence and abstract thought associated with the human species. Cultural aspects of human evolution are discussed in Chapter 11.

## Post-Cranial Skeleton

Darwin reasoned that the first step in the evolution of humans from apelike ancestors was the development of erect posture. Modification of the hind limbs for support and locomotion on the ground freed the forelimbs to carry food and manipulate objects—leading to the habits of tool using and tool making. As man's earliest ancestors were probably tree dwellers, transition to ground living required modification of arboreal traits. Comparison of the skeletons of modern ape and man details the changes that occurred as the apelike ancestors of man left the forests and moved onto the grassy savannahs of the Pliocene Epoch.

In arboreal animals the long bones of the forelimb and hindlimb are about the same length and are gently curved, while in man the forelimb is shortened and the bones of both limbs are straightened (see Figure 6.1). In the human foot the "great toe" occurs in line with the other four digits, for balance and pushing off when walking. The "heel bone" of man is enlarged and provides a landing point in walking and aids in balance—which is also enhanced by the curve of the human spine as opposed to the straight spine of the ape. Finally, the pelvic girdle of man is short and widely flared, while in the apes it is elongated and projects up the spine.

## The Skull

The skull encloses the brain of the vertebrate animal and provides openings for the sensory apparatus. The jaws and teeth are adapted to feeding habits and may also function as weapons of offense and defense. Thus, the skull provides a wealth of information about the life-style and behavior of an animal.

In general, ground dwellers have a coarser diet than tree dwellers and this difference is reflected in the shape of the teeth and the way they become worn during the life of the animal. Human teeth are smaller than those of the ape, and the total number of teeth is less (Figure 10.3). The

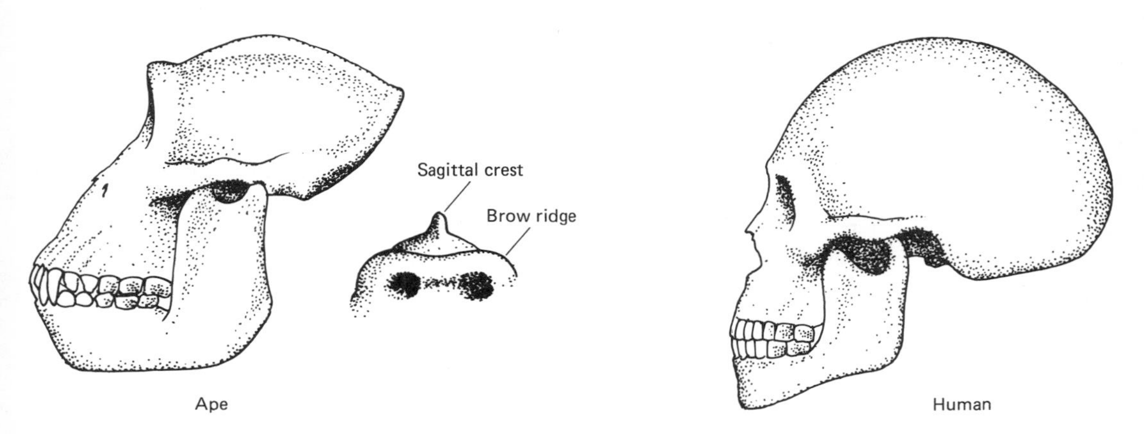


**Figure 10.3** Comparison of human and ape skulls.

tendency toward reduction in tooth size is especially obvious in the canines. In predatory animals, or animals that use their teeth for fighting, the canine teeth are large and pointed. In human evolution, the forelimb and the tool became the weapons used for these functions and the canine teeth became shortened and little different from the premolars. Reduction of the muzzle and the flat profile of the human skull result in part from this trend.

As the teeth became smaller so did the lower jaw and the muscles that activate it. In the ape, which retains the fighting function of the mouth, the jaw muscles are so large and powerful that they require a special structure, the *sagittal crest,* to serve as a point of attachment (Figure 10.3). The sagittal crest is absent from the earliest stages of human evolution. Another bony structure, the *brow ridge,* also serves in muscle attachment and is greatly reduced or absent in humans.

Finally, the human cranium is larger than that of the apes. The cranial capacity of modern man ranges from 1200 to 1500 cubic centimeters (cc) while that of the ape is 400 to 500 cc. The enlargement of the skull to accommodate the larger human brain has produced the elevation of the forehead.

The above array of physical attributes makes it easy to distinguish the skeleton of modern man from modern ape. However, fragmentary remains of fossil animals often display a mixture of apelike and human characteristics that tax the ability of the scientist to make a definitive decision as to whether the animal in life was more like an ape or a human.

## Human Origins

The fossil record of apes and humans is meager and interpretation of fragmentary skeletal remains has yielded many heated debates concerning their classification. And what we know—or think we know—about human evolution is being constantly revised as new discoveries are made and as new interpretations are given to existing fossils. What is presented here is an outline of current thought concerning human evolution and will no doubt be modified within a few years.

The oldest known fossil ape, *Pliopithecus,* dates to the Fayum strata of Egypt. From this center of distribution apes spread over Africa, Europe, and Asia. By the Miocene Epoch several genera of apes had evolved, collectively referred to as the dryopithecines. These apes persisted for over 15 million years and displayed a great range of attributes—from the primitive *Aegyptopithecus* and *Dryopithecus,* with their enlarged canines and pronounced muzzles, to the more rounded profile of *Proconsul* (Figure 10.4). From this large and diverse group descended lines leading to modern apes and *hominids* (humans and their ancestors).

The first fossil with significantly human characteristics was a creature called *Ramapithecus,* who lived in Africa 14 million years ago. Only skull fragments and teeth of *Ramapithecus* have been found and so nothing is known about its posture, brain size, or culture—but its dental pattern was more human than apelike, and the wear pattern of its teeth suggest that its

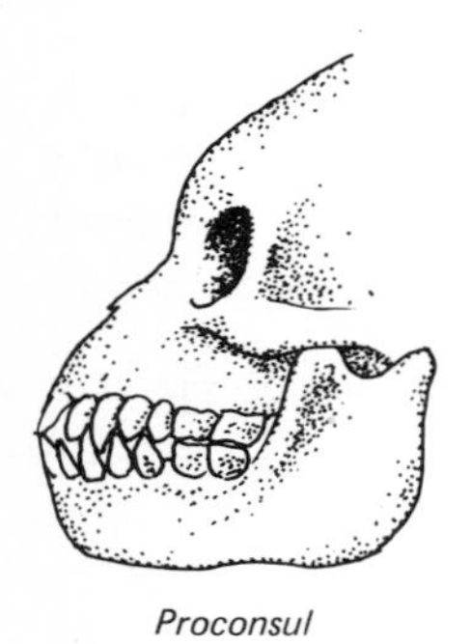

*Proconsul*

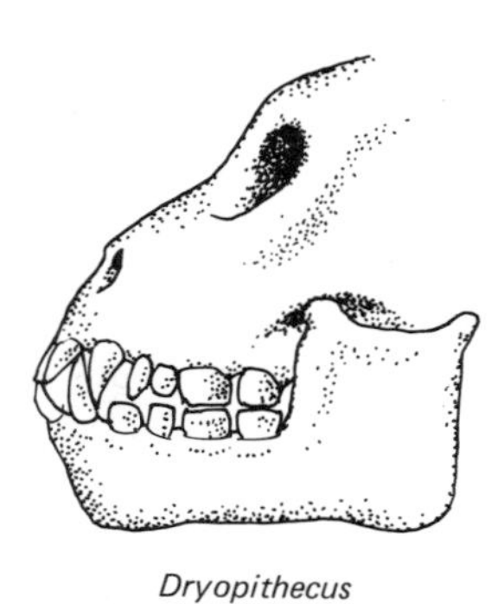

*Dryopithecus*

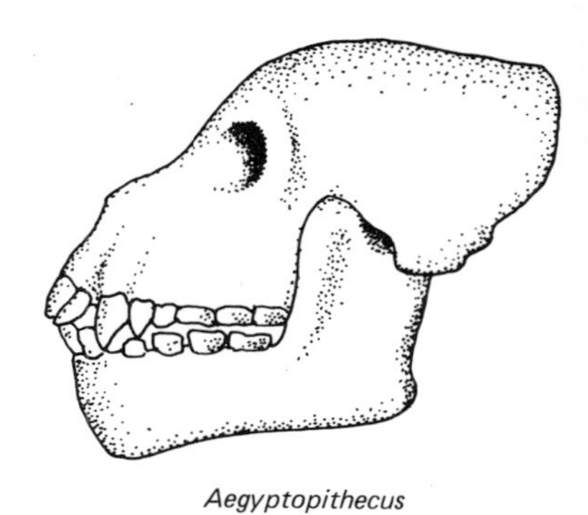

*Aegyptopithecus*

**Figure 10.4** Representative dryopithecine apes. *Aegyptopithecus* lived about 30 million years ago; *Dryopithecus* about 25 million years ago; and *Proconsul* about 20 million years ago.

diet consisted of the coarser foods associated with the ground dweller. *Ramapithecus* was not human, but it is generally agreed that it represents the earliest known step in the divergence of the human line from that of the apes.

Following *Ramapithecus* there is a long gap in the fossil record of human descent until about 3 to 4 million years ago when several groups of humanlike creatures appear. Quite possibly all had a common ancestor in *Ramapithecus,* but the gap in the fossil record makes this a tentative conclusion.

## The Australopithecines

The australopithecines were a diverse group of primates that occupied eastern and southern Africa three to four million years ago. Although experts do not agree as to the exact number of species that existed, all can be grouped in the genus *Australopithecus* (southern ape–man) and, for our purposes, we need consider only two species: *Australopithecus robustus* and *Australopithecus africanus.*

*Australopithecus robustus* was a stocky hominid just over five feet tall and powerfully built. Although a woodland dweller, his skeleton shows that he was capable of nearly erect posture and so must have spent much of his time on the ground. The teeth of *Australopithecus robustus* indicate that he was primarily a vegetarian although he may well have supplemented his diet with small game. His brain size was small (400-500 cc) and he had an apelike skull with a sagittal crest and pronounced brow ridges (Figure 10.5).

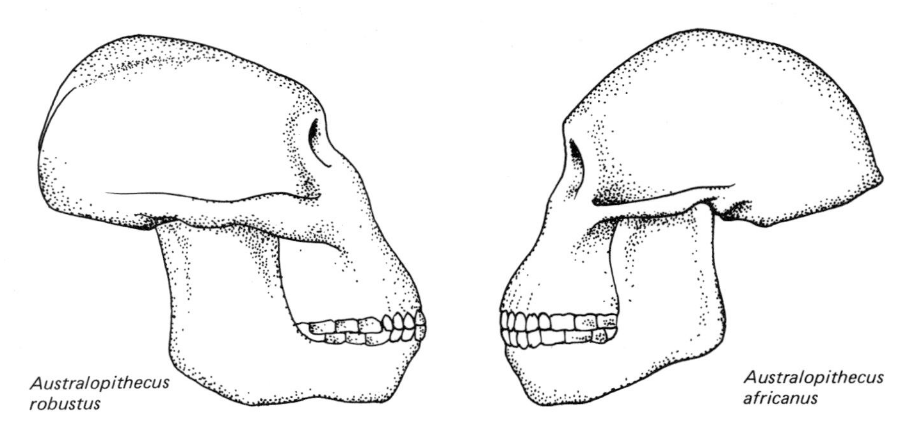


**Figure 10.5** Australopithecines.

*Australopithecus africanus* was a smaller, more graceful creature, about four feet tall and capable of fully erect posture. His skull was more humanlike in that it lacked a sagittal crest and the greatly protruding brow ridges of *Australopithecus robustus* (Figure 10.5). The brain capacity of the gracile form was the same as that of *Australopithecus robustus* but, with its smaller body-size to brain ratio, *Australopithecus africanus* was the more intelligent of the two. *Australopithecus africanus* inhabited the grassy savannahs of eastern Africa and lived the life of a small game hunter. Quite possibly, *Australopithecus robustus* and *Australopithecus africanus* occupied the same relative ecological niches as are occupied by the gorilla and native African tribesman of today.

The australopithecines, and especially *Australopithecus africanus*, closely approached being human—their posture was essentially erect and *A. africanus* had attained the ability to use and make crude tools. But, because of their small brain size and primitive skull features they are not included in the family of man. The australopithecines appear to be a terminal line that became extinct some 800,000 years ago.

## Other Hominids

In 1961 Dr. L. S. B. Leakey found the fossil remains of a 2 million year old hominid creature that he subsequently named *Homo habilis*, placing it in the family of man. Since that time *Homo habilis* has created a great deal of controversy between those who believe it was truly the first member of the genus of man and those who believe that *Homo habilis* is simply another member of the genus *Australopithecus*. The brain capacity of *Homo habilis* is larger than other australopithecines (670 cc), it had erect posture, the hand bones are humanlike, and it both made and used tools. The major question to be answered is whether the characteristics of *Homo habilis* are distinct enough to place it in a separate genus from the australopithecines or whether they fall within the normal range of variation of the genus *Australopithecus*.

Another find that confuses the history of human evolution was made by Donald Johanson, of Case Western Reserve University, in the Afar region of Ethiopia. Although not completely classified, Johanson believes that the Afar skeletons may belong to the genus *Homo*. The skull remains are fragmentary but the hands of these hominids were very humanlike and the femur shows that they walked erect. If these fossils are ultimately classified as members of the genus *Homo* the antiquity of man will have been pushed back even further, because the Afar hominids occur in rock strata dating to over 3.0 million years, and thus these hominids must have begun their divergence at an even earlier date.

Perhaps all of these manlike creatures trace their lineage directly to *Ramapithecus,* but we have no record of the events that occurred between the time of *Ramapithecus* (14 million years ago) and the advent of these hominids more than 10 million years later. Quite possibly several hominid lines did diverge from ramapithecine stock and underwent parallel evolution—i.e., they lived in similar habitats and underwent similar but not identical adaptive changes.

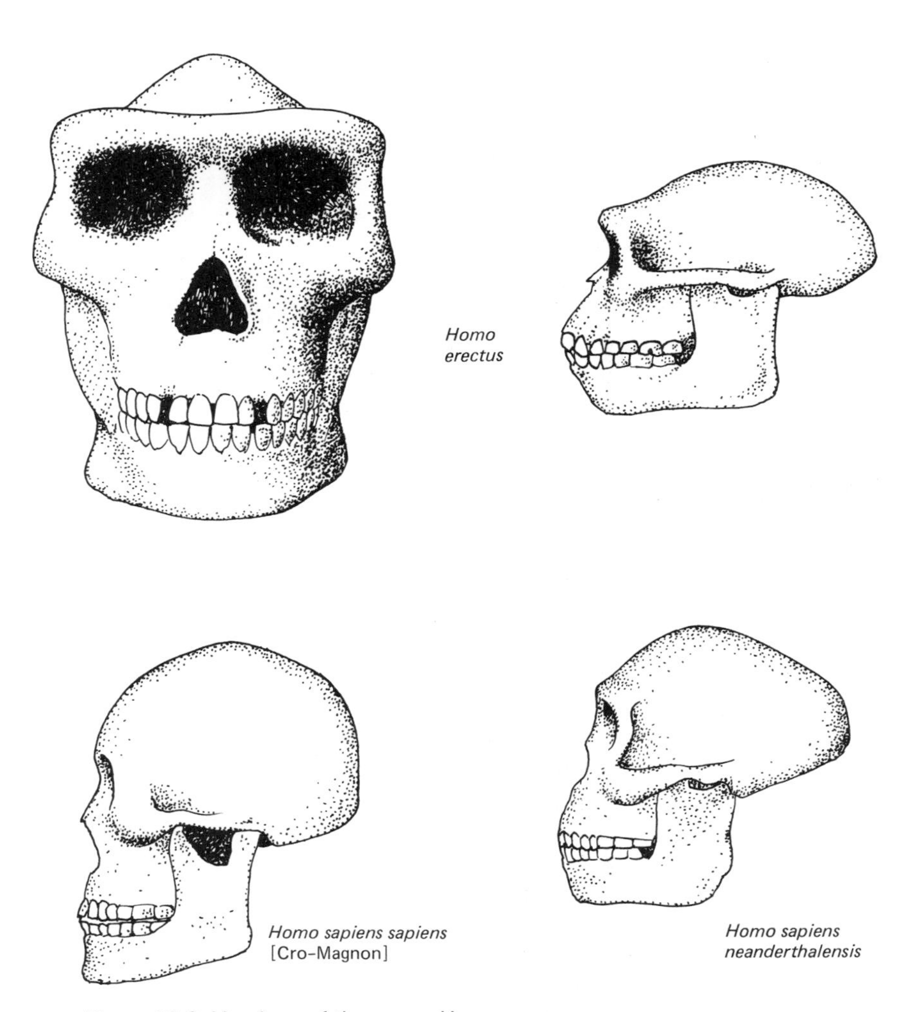


**Figure 10.6** Members of the genus *Homo.*

## Homo Erectus

*Homo erectus* is the first generally accepted member of the genus of man. He had erect posture and a brain capacity that ranged from 800–1300 cc but, because his average brain capacity remained less than that of modern man (1200–1500 cc) and because of persistent primitive skull features, he has been assigned to the species *erectus* rather than *sapiens* (modern man). Fossil remains of *Homo erectus* have been found in China, Java, and Africa—the familiar "Java man" and "Peking man" belong to this group (Figure 10.6). Over the broad geographic range of *Homo erectus* there appeared a wide variation in physical attributes, at least as great as occurs in the races of modern man. Some experts believe that the races of modern humans have evolved directly from the geographic races of *Homo erectus,* whereas others think that racial divergence has occurred much more recently, from a common stock of man that arose within the last 35,000 to 40,000 years.

The attributes of *Homo erectus* definitely qualify him as human—from his erect posture and large brain to his advanced tool-making abilities and big-game hunting culture. He inhabited the Earth from approximately 1.5 to 2 million years ago until his disappearance 200,000 to 300,000 years ago. Following the extinction of *Homo erectus* another gap in the fossil record occurs. Then, about 90,000 years ago, a new species of man appeared, *Homo sapiens neanderthalensis.*

## Neanderthal Man

*Homo sapiens neanderthalensis,* like *Homo erectus,* is a classificatory group that includes a variety of fossil types that have been found over a wide geographic range. The Neanderthals existed from roughly 90,000 years ago until 40,000 years ago and showed both temporal and geographic variation. Physically, Neanderthals were erect-postured with a brain capacity that generally ranged from 1200-1500 cc although some exceeded 1800 cc. However, because Neanderthals lacked a chin and had large, heavy jaws and teeth, they are assigned to the subspecies *neanderthalensis* whereas modern man is classified as *Homo sapiens sapiens.*

A summary of the geographic and chronological distribution of Neanderthals is presented in Figure 10.7. If it is assumed that all Neanderthals descended from a single stock, Heidelberg man is the type from which all others diverged and differences among the various races of Neanderthals arose from the geographic isolation of subpopulations as they migrated through Europe, Africa, and the Middle East into Asia. The classic Neanderthal type, pictured as a squat, beetle-browed individual, was probably a terminal line of a subpopulation highly adapted to the cold

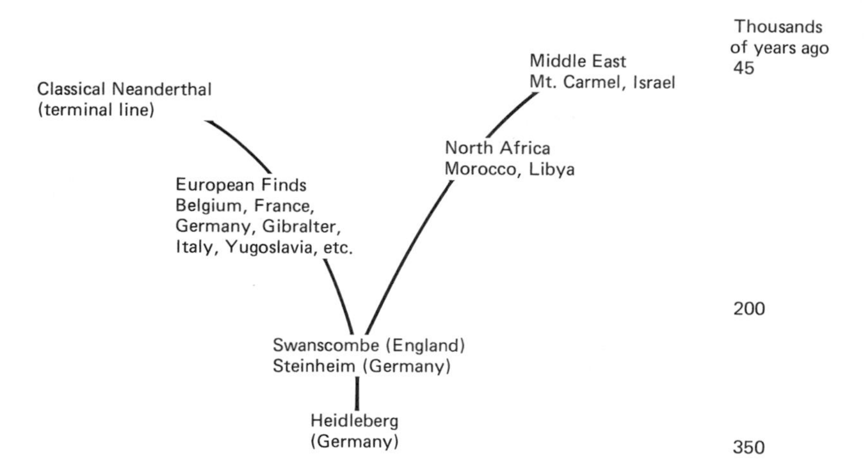


**Figure 10.7** Distribution of *Homo sapiens neanderthalensis*. The oldest Neanderthal find comes from Heidelberg, Germany. One line of Neanderthals ranged over Europe and another developed in North Africa and the Middle East.

glacial climate that gripped Europe during the Pleistocene Epoch. This race became extinct about 40,000 years ago. Another group migrated southward to Africa and its fossils have been found in Rhodesia, North Africa, and the Middle East. Some were interred in caves in the Mt. Carmel region of Israel and provide circumstantial evidence that Mt. Carmel man was the ancestor of modern man, *Homo sapiens sapiens.*

The remains of Mt. Carmel man have been found in cave deposits that date to nearly 45,000 years ago. In younger layers of the same caves occur an enigmatic mixture of fossil types and cultural elements (tools, pottery, etc.) that are transitional between Mt. Carmel man and modern man. In still younger layers appear the remains of *Homo sapiens sapiens.* This fossil sequence may represent an evolutionary pattern between Mt. Carmel man and modern man but it is also possible that Mt. Carmel man, instead of evolving into modern man, was exterminated by *Homo sapiens sapiens* when he immigrated into the Mt. Carmel region from some other center of origin. If this is the case, *Homo sapiens sapiens* may be a direct descendant of *Homo erectus* and have evolved parallel to the Neanderthal lines (Figure 10.8).

## Homo sapiens sapiens

The richest finds of early modern man come from the Mt. Carmel area of the Middle East and from southern France. Cro-Magnon man (Figure 10.6) is typical of these finds—he was completely modern with a large

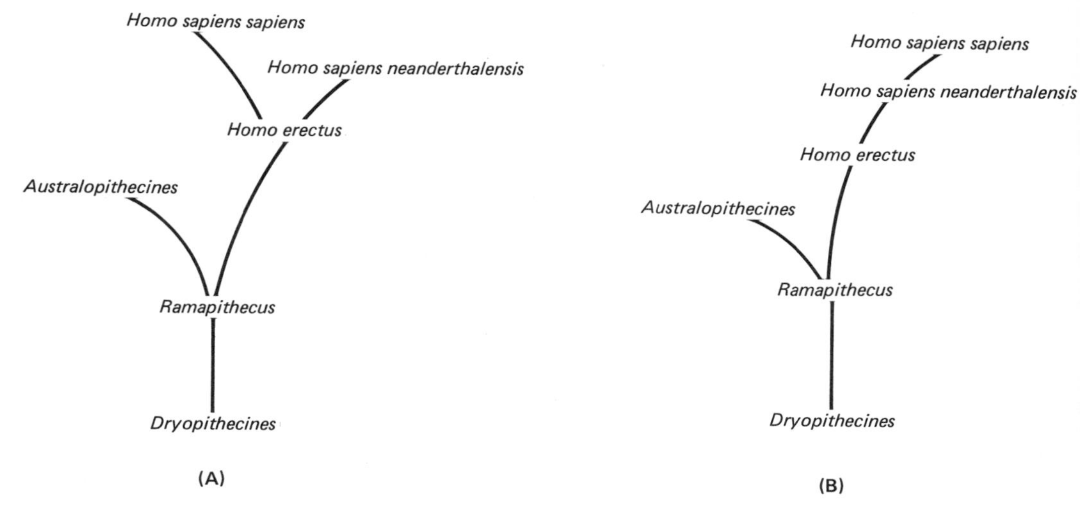


**Figure 10.8** Possible patterns of human evolution. Modern man, *Homo sapiens sapiens*, may have evolved directly from *Homo erectus*, as in (A); or from Neanderthal stock, as in (B).

brain, high forehead, prominent chin, and tall stature. From the time of his appearance, about 35,000 years ago, there has been essentially no change in the physical attributes or intellectual ability of the human race. The major change during this period of time has been the advance of human culture and civilization.

# Human Evolution Two

## Introduction

Social behavior, in the sense of cooperative effort among members of a population, is found in many animal species and ranges from the highly structured societies of ants and bees to the more loosely organized, amorphous flocking of birds. While the social behavior of lower animals, such as insects, is instinctive, learning is a major component of social behavior among the anthropoids and especially humans. A common factor in all social behavior is that it increases the likelihood for survival of the species and thus has a positive selection pressure.

Little in human social behavior is unique, any more than there are uniquely human physical traits. However, if there is a distinctly human quality it is that human social behavior has progressed into cultural development. Social behavior is group behavior as opposed to individual activity. Culture is the total knowledge of a society and includes all phases of learning: religion, art, philosophy, science, technology, and folk wisdom. Only man appears to be conscious of his culture and is aware of his history.

Tracing man's social–cultural heritage is difficult because of many gaps in the fossil record and because fossilized skeletal remains, in themselves, tell us little or nothing about the social activities of the organism in life. The social history of the hominid line has been interpreted largely from tools, burials, building sites, pottery, and art. Unfortunately, such artifacts are often either missing or poorly preserved—especially those dating to the earliest stages of hominid evolution.

## Primate Behavior

The study of primate behavior provides insights into the early stages of hominid social evolution. For example, macaques and chimps are capable of walking for short distances on their hind legs—a habit that frees the forelimbs for gathering and carrying food and elevates the head and sensory apparatus enabling them to see over shrubs and tall grass to detect predators. Presumably, *Ramapithecus* was capable of similar posture, and because the trait conferred an adaptive advantage in the terrestrial habitat, natural selection directed the course of hominid evolution through the series of changes that led to the development of truly erect posture (see Chapter 10).

The fact that some primates eat meat is also of interest to students of human evolution. Gorillas occasionally organize impromptu hunting parties to supplement their normally vegetarian diet with small animals, including monkeys. Early hominids turned to the hunting of small animals for an additional food source due to great competition for the sparse vegetation found on the ground. Their success as small-game hunters was dependent upon formation of small hunting bands that cooperated in stalking prey—and shared the spoils of the kill. This was an important step in hominid evolution because this kind of group activity placed great emphasis on cooperative behavior and the development of communication skills.

## Tools and Culture

The advent of tool using and tool making—a common practice among many kinds of animals—greatly enhanced the success of the hunting bands. Gorillas select twigs, carefully remove the leaves, and thrust them into termite holes. When termites crawl onto the twigs, the apes have obtained choice morsels to eat. This activity by the gorilla includes both tool making and tool using. The manipulative ability of the primate's forelimb is partly responsible for such activity, but so is the ability to reason—to recognize the advantage of a particular activity and to remember and repeat that activity (learning).

A major advance in tool making came when hominids learned to use stones and sticks as weapons of offense and defense. Chimps have been observed to attack a stuffed tiger with stones and stout sticks, becoming more and more aggressive when the enemy remains inert. Hominids learned the advantage of such tools and made them part of their culture. As hominids gained sophistication in shaping inanimate objects to meet specific needs, bone, stone, and fire became adjuncts to human activities. Man learned to fashion metals into jewelry, knives, and plowshares—and to use dyes to color his clothes, pottery, and works of art (Figure 11.1).

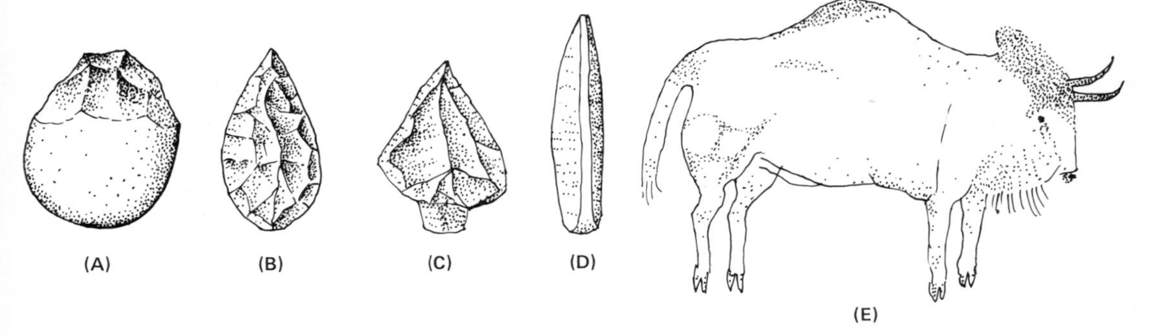


**Figure 11.1** Cultural elements of human evolution. (A) Pebble tool dating to over 500,000 years ago. (B, C, and D) Tools from the Stone Age 150,000 to 30,000 years ago. (E) Cave art dating from 15,000 to 20,000 years ago.

As tool making became a part of hominid culture it also became an influence in hominid evolution. Those individuals with the greatest manipulative ability of the hands were the most adept tool makers and thus had an advantage in the competition for survival. Because manual dexterity became a factor in natural selection, the human hand has evolved the capability of more precise and delicate operations than has the "hand" of any other animal.

## Intelligence

Intelligence is another important factor in the success of a tool-oriented culture. All of the technological capabilities of the human race are learned activities, and as learning is a mental activity, natural selection favored the more intelligent individuals of a population. It is a sobering thought that all of the great literary, artistic, medical, and technological advancements of modern culture would be completely lost if a single generation were born and reared without any teaching or record of their past.

Increased brain size in the hominid line is an extension of a general trend in primate evolution. Survival in the arboreal habitat required greater use of the forelimbs, which in turn led to increase in the size of the brain centers that control them. In the hominid line, advances in tool using and tool making further accentuated the development of these brain centers and it seems likely, since learning is important in such activities, selection also favored the more intelligent members of the population.

## Stages of Development

Paleontological evidence supports the belief that early hominids were hunters. Johanson's Afar hominids lived in small groups whose diet included small game. Australopithecines both used and made primitive hunting tools. The fact that both groups established temporary dwelling sites indicates they had attained some level of social stability.

*Homo erectus* was an advanced tool maker and his semipermanent habitations contained hearths—clear evidence that he had mastered the use of fire. The Neanderthals fashioned still finer tools and were hunters of big game—mastodon, bison, the wooly rhinoceros, bear, and deer. The Neanderthals were cave dwellers whose dawning religious awareness was expressed in their burial of the dead.

Cro-Magnon man was at once a big game hunter, artist, and religious being. The cave art left by these men reflects an abstraction of thought and a desire to communicate man's inner experience through visual communication, not before evidenced.

## Language

One facet of human cultural development not interpretable from paleontological evidence is the origin of language. Communication by sound is found among many animals, from the high-pitched "beeps" of porpoises to the chatters, howls, and grunts of monkeys and apes. Chimps can correlate word symbols with their physical representations—such as recognition symbols that stand for food, water, or some other chimp (Figure 11.2). But man is one of the few animals physically capable of forming words and is the only animal who has developed complex spoken and written languages. Speech facilitates the communication so important to social living and also the expression of abstract concepts of emotion, religion,

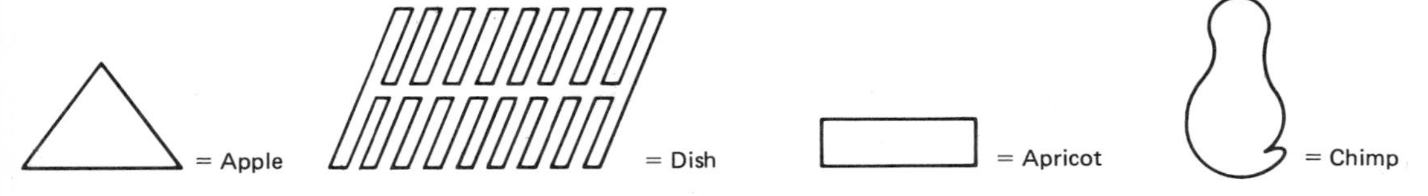


**Figure 11.2** Chimp recognition symbols.

art, and philosophy. The word, it is said, is the unit of cultural transmission, just as the gene is the unit of biological transmission.

## The Rise of Civilization

So long as man remained a hunter he could not establish permanent settlements, for he had to follow the wandering herds that were the center of his activities. But, at some recent stage in his history he made the transition from hunter to agriculturalist, an event that directed his efforts toward survival within a new framework. The techniques of plant and animal breeding, of tilling and fertilizing the soil, and of developing the tools needed to eke out a living from the land occupied his time and interests. The required tasks were sedentary and, as man became a farmer, he took his first step toward the development of permanent settlements and the establishment of great civilizations.

The oldest civilizations appeared in the Middle East (Figure 11.3), in the Indus Valley, in the "fertile crescent" of the Tigris and Euphrates Rivers, and along the banks of the Nile. When man first came to the area that is now Egypt the region received far more rainfall than occurs today, and the land that is now desert was then a marshy woodland abounding in game. The inhabitants of the area were hunters who pursued species no longer found in the arid wastes of the Egyptian deserts. The modern deserts were produced by a drying climate that drove out the forest animals and restricted life to a narrow fertile band bordering the life-giving river. With the elimination of most game species, the people of the Nile Valley became increasingly reliant on domesticated animals and plant food and they learned methods to increase the abundance of those plants that meant the balance between life and starvation. The success of the early Egyptian agriculturalists is evidenced by the wealth of their civilization and the magnitude of their architecture. The building of the pyramids and temples of ancient Egypt could not have been accomplished by a nomadic tribe without centralized authority, social stability, and, most likely, some

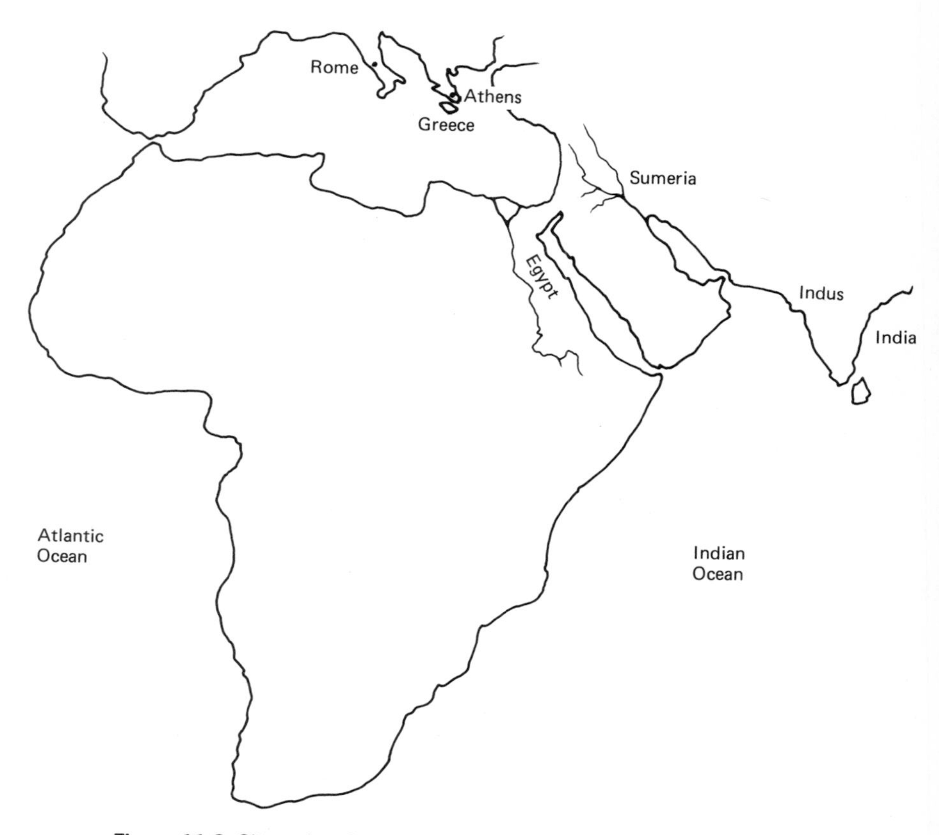


**Figure 11.3** Sites of early civilizations.

form of bureaucracy. Perhaps more than anything else, the monuments built by man reflect the levels of intelligence, planning, communication, organization, and stability of his societies.

The transition from hunter to agriculturalist occurred in many human populations in many parts of the world, with many variations. In areas where the land was poor, people remained nomadic hunters, whereas in marginal areas transient agricultural societies developed. Where the land was rich, man developed stable communities, and as he prospered, trading centers grew into cities and cities into advanced civilizations where artisans and artists flourished, as did the priesthood, philosophy, and science. Given relative abundance and leisure, man addressed his thoughts to the fundamental questions that seem to have always been with him—contemplated by Neanderthal man as he buried his dead and by Cro-Magnon man as he worked his rich colors onto the walls of the caves of Southern

France—"Where has man come from?", "What is his place in the universe?", "What is his future?", and "What is his relationship to God?"

## Culture and Evolution

In the latter part of the eighteenth century, interest in man's cultural heritage was stimulated by tales brought back by seamen and travelers who had voyaged to the far corners of the Earth and had seen cultures ranging from the primitive tribes of Tierra del Fuego to the plush feudalism of the Orient. This interest led to the inevitable question of why such a great cultural disparity should exist among contemporary peoples.

One answer to this question was found in application of the principles of organic evolution to human cultural development. Organic evolution has taken place as a series of continuous steps—a species arises, exists for a time, and is supplanted by a new, better adapted species. No step in the sequence is skipped, none omitted. Cultures, it was reasoned, evolve in the same way—beginning at a primitive stage and passing through a series of increasingly sophisticated steps until they ultimately arrived at the level of perfection displayed by the refined European of the 1800s. This *unilineal theory* of cultural evolution appeared in several forms; one, proposed by T. H. Morgan, stated that all human societies must pass through three stages: savagery, barbarism, and civilization. A tribe at the level of savagery must complete the intervening step of barbarism before it can become civilized.

Support for the unilineal concept was widespread—with unfortunate results. The strong implication of this theory is that societies that had not achieved the cultural level of nineteenth-century Europe were inferior culturally and biologically. A movement called Social Darwinism grew out of this belief, a movement that was not Darwinian in any sense. The Social Darwinians propounded a theory of biological racism—the idea that culturally less advanced peoples were inferior to members of advanced societies. These ideas were used by apologists for the colonial movement and the excesses of the trust and cartel founders. "Survival of the fittest" justified the belief that the more advanced and better adapted races of man should control those at lower stages of evolution. Hitler and his Nazi theoreticians saw justification for their concept of an Aryan race and world conquest in these arguments for racial superiority.

Fortunately, it is now realized that the reasoning behind the unilineal theory and Social Darwinism is erroneous. All races of humans are biologically equivalent, racial variations being minor at the genetic level. Differences in culture reflect variations in the physical environment in which people live as well as their history. The isolated bushman of the Kalahari

Desert of South Africa, expending the bulk of his energies in the task of scraping out a meager existence from the harsh desert environment, could hardly be expected to develop a technological society from the resources at his disposal. But the important point is that the bushman can learn the skills of a technological society. Culture is learned, not inherited. The unit of cultural transfer is the word and not the gene. The son of a nuclear physicist, raised from infancy among Berber tribesmen, will have no concept of the atom or the cyclotron; but he will know the language, art, religion, and mores of the Berber culture. And the son of the Berber can learn to be a computer programmer or to fly a jet plane. This potential for *cross-cultural transfer* provides the fallacy in the argument for the unilineal theory or biological racism. Unlike biological traits, culture can be transmitted from tribe to tribe and from society to society.

## Man's Evolutionary Future

The evolutionary status of modern man is the subject of a great deal of speculation and controversy. Some believe that man has so gained control of his environment that he no longer evolves biologically and that his only change in the future will be cultural. This position is difficult to accept when it is considered that the essence of evolution is: (1) biological (genetic) variation; (2) differential reproduction; and (3) environmental variation. Certainly there is not genetic uniformity in the human species. In fact, the increased levels of high-energy radiation and potentially mutagenic chemicals in man's environment must work to increase mutation rates and thus increase genetic variability in the human gene pool. Since these mutations are, for the most part, deleterious, this is a sobering thought.

For most human societies the factors in differential reproduction are different today from those experienced by man of 10,000 or even 1000 years ago. Few humans now face death from predators, and advances in medicine have reduced the incidence of death from disease or injury. Still, not all the world is well fed, and the peoples of many regions teeter on the brink of agricultural disaster. And such accoutrements of modern society as stress and stress-related disease may be the twentieth-century counterpart of infectious disease in affecting the life span and reproductive potential of susceptible individuals. Also, the "pill," abortion, and increased social acceptance of the nonmarried status are having their effect on the population geneticist's ideal of completely random mating. In addition to these factors, sociological studies indicate that people tend to marry within racial, economic, and IQ groups. Thus, human populations show a strong tendency toward differential breeding.

Changes in the human environment are no less evident than are the existence of genetic variation and differential reproduction. Stress was mentioned above, as was the increased levels of radiation and chemicals in man's environment. The entire urban scene is a relatively new habitat for man—given his more than four million years of evolution—and man-made changes in the atmosphere we breathe and the water we drink certainly represent environmental variations. Further, beyond the scope of man's ability to control his environment, there remains the possibility of great climatic changes such as have occurred many times in the geologic history of the Earth.

## The Classical Hypothesis

What is not clear is the *course* of human evolution in the future. Two schools of thought have developed on this subject, with completely contrasting points of view. The *Classical Hypothesis* states that the genetic future of the human race is rather dim. An increasing mutation rate is saturating the human gene pool with detrimental genes and, in addition to this, medical science is contributing to the genetic load through treatment of genetic birth defects that formerly caused death or debilitation before the afflicted individual reached reproductive age. With treatment, many afflicted individuals now live to produce offspring and pass on detrimental genes that once would have been eliminated from the gene pool.

Some believe that selective mating patterns lower the vitality of the human gene pool because couples in higher IQ groups tend to have fewer children than those in lower IQ groups, thus causing a generation by generation decrease in the proportion of high-IQ individuals in the population.

## The Balance Hypothesis

Opposed to the Classical Hypothesis is the *Balance Hypothesis* of human genetics. The latter grants the general observations of the Classical Hypothesis but disputes its conclusions. A major point cited by the followers of the Balance Hypothesis is that evolution favors populations with genetic diversity. Too much uniformity in a gene pool diminishes flexibility, and highly inbred gene pools are unable to respond to environmental change. Another argument in favor of gene-pool diversity is that *hybrid* plants and animals tend to be more vigorous and hardy than highly inbred varieties.

The problem of genetic load is a real one, but the supporters of the Balance Hypothesis point out that genes are neither 100% "good" nor

100% "bad." A gene that is neutral or detrimental in one environment may actually have a beneficial effect under a different set of environmental conditions. A gene conferring antibiotic resistance in bacteria has no known function in the absence of streptomycin, but when that chemical is present the mutant gene is essential to survival of the organism. Similarly, the mutant gene that causes the "sickling" condition of hemoglobin actually has a positive survival value in malarial regions. As to selective breeding within IQ groups, it is argued that geniuses are as likely to be born of parents of normal IQ as those of high IQ.

The proponents of the Balance Hypothesis also look to the relatively new field of "genetic engineering," in hopes that it may soon be possible to alter genes chemically so that potentially harmful genes may be converted to forms that do not produce genetic defects. This technique could, of course, eliminate or greatly diminish the problem of genetic load. Those who endorse the value of genetic engineering foresee the day when such techniques might go beyond the treatment of genetic defects and be extended to the alteration of human heredity for the betterment of the human race—a controlled evolution. Should the capability to perform such feats become possible, it will raise moral questions as to what the "ideal human" or "ideal society" should be. Such questions exceed the limitations of scientific methodology and, given the inability of science to function authoritatively in the areas of morality, philosophy, and religion, it seems that the answers to such questions must come from a broad base and not solely from the scientific community.

# Glossary

**Abiotic** Not biological. Not of biological origin.

**Absolute age** Time before the present determined by radiometric dating.

**Acellular** Not composed of cells.

**Adenosine triphosphate (ATP)** A chemical compound that stores chemically bound energy.

**Allantois** Saclike structure in the amniote egg that receives waste products produced by the developing embryo.

**Allele** Any one of the genes possible at a single locus.

**Amino acid** A unit of protein synthesis. There are some twenty-odd kinds of amino acids that make up the proteins of all living organisms.

**Amnion** Saclike structure that surrounds the embryo in the amniote egg; it contains a watery fluid that protects the embryo and keeps it moist.

**Analogy** Organs that are similar in function but have a different genetic origin. Such organs are structurally different.

**Antheridium** Botany: a male sex organ that contains the sperm cell.

**Archegonium** Botany: a female sex organ in which the egg is surrounded by sterile cells.

**Biotic potential** Maximum reproductive capacity of an organism under ideal environmental conditions.

**Carnivore** An animal that feeds predominantly or entirely on other animals.

**Carrier** A heterozygous individual who has a potentially detrimental recessive allele.

**Cartilage** A kind of connective tissue found in all vertebrates.

**Catalyst** A substance that alters the rate of a chemical reaction without being altered by the reaction.

**Chlorophyll** The green pigment that absorbs light energy in the process of photosynthesis.

**Chromosome** The rodlike structure found in the nucleus of cells; it is made up of genetic material (DNA) and protein.

**Coldblooded** Said of an animal whose body temperature fluctuates with that of the ambient environment (e.g., fish, amphibians, and reptiles).

**Colony** An aggregation of cells that grow in association with one another but among which there is little or no cellular specialization.

**Cytochrome** One of a group of enzymes that mediates the process of cellular respiration.

**Cytoplasm** All of the living substance of a cell except the nucleus.

**Deoxyribonucleic acid (DNA)** Chemical component of which genes are made.

**Diploid** A nucleus that contains two sets of chromosomes.

**Divergence** The process in which natural selection produces offspring that vary genetically from the parents.

**Dominant** A gene that controls phenotypic expression regardless of the nature of the second allele.

**Embryo** Early stage in the development of a new individual of the species. The embryonic stage commences with the first division of the zygote.

**Enzyme** A protein that functions as a catalyst in cellular reactions.

**Erosion** Includes the geologic processes of: (1) breakdown of rock material (weathering); (2) transport of rock material by water, wind, or ice; and (3) deposition of transported material.

**Eucaryotic cell** A cell that has (1) a distinct nucleus, separated from the cytoplasm by a membrane, and (2) cytoplasmic specialization.

**Extant** Currently living plant or animal species.

**Extinct** Plant and animal forms that no longer exist.

**Fauna** All animals present in a specific geographic region or a specific geologic period.

**Fertilization** Union of the nuclei of sperm and egg cells.

**Flora** All plants present in a specific geographic area or a specific geologic period.

**Fossil** Any trace of formerly existing life forms including (1) preserved specimens, (2) petrified specimens, and (3) casts, molds, or footprints.

**Fossil assemblage** A group of plant and animal species typical of a particular geologic time.

**Fossil correlation** The geologic principle that states that rocks that contain a similar fossil assemblage must be of the same geologic age. A relative dating technique.

**Gametes** Sex cells (sperm and egg); haploid cells produced by meiotic cell division.

**Gametophyte** In alternation of generations: the haploid plant that produces gametes by mitosis.

**Gene** The unit of hereditary transmission; a series of nucleotide units making up part of a chromosome.

**Gene flow** The movement of genes from one population to another as a result of migration of individual organisms.

**Gene frequency** The proportion of alleles in the gene pool.

**Gene pool** All genes present in a specific population.

**Gene recombination** (*See* Recombination).

**Genetic drift** Changes in gene frequencies due to chance events; most influential in small populations.

**Genetic load** The decreased "fitness" of a gene pool that results from the occurrence of detrimental alleles.

**Genotype** The genes actually present in an organism.

**Geographic barrier** A physical barrier (e.g., river, mountain range, or ocean) that prevents interbreeding between subpopulations of a species.

**Geographic isolation** Division of a species population into subpopulations as a result of (1) the establishment of geographic barriers, or (2) spatial isolation.

**Geographic Race** (*See* Race).

**Half-life** The time required for one-half of the radioactive nuclei present to decay.

**Haploid** Having only one chromosome of a pair; typical chromosome number of the gametic stage of all organisms.

**Herbivore** An animal that feeds predominantly or entirely on plant material.

**Heterozygous** Having two different alleles of a gene at a specific locus.

**Hominids** Humans and their close relatives.

**Homology** Similarity in structure that has arisen from common genetic ancestry.

**Homozygous** Having two identical alleles of a gene at a specific locus.

**Hybrid** The offspring of a cross between genetically different parents.

**Industrial melanism** The situation in which darkening of the natural habitat by air pollutants has resulted in the selection of dark (melanic) forms of an insect species.

**Inorganic evolution** Origin of organic molecules from nonliving matter.

**Insectivore** An animal that feeds on insects.

**Intraspecific competition** Competition for resources among members of a single species.

**Invertebrates** Animals having neither backbone nor internal skeleton.

**Isolating mechanism** A barrier (ecological, physical, or behavioral) that prevents successful reproduction between individuals of two different species.

**Lack of dominance** A genetic relationship in which the heterozygous individual has a trait expression intermediate to that produced by the homozygous dominant or homozygous recessive genotypes.

**Locus** Position a gene occupies in a chromosome.

**Lysosome** A subcellular unit found in the cytoplasm of eucaryotic cells. It stores digestive enzymes.

**Meiosis** A special kind of cell division in which the chromosome number of the parent cell is reduced by one-half. The cells produced by meiosis are called gametes.

**Melanic** Dark-colored.

**Melanin** A dark-colored pigment.

**Mitachondrion** A subcellular unit found in eucaryotic cells. It is the site of cellular respiration.

**Mitosis** A kind of cell division in which daughter cells have the same chromosome content as the parent cell.

**Monophyletic model** A phylogenetic model in which all species are traceable to a common ancestor.

**Multiple alleles** More than two allelic forms of the gene at a single locus.

**Mutation** Any heritable change in a gene.

**Notochord** The strong, flexible rod that serves as a supporting structure in the sea lancelet and which appears during the embryonic stage of more advanced vertebrates. Replaced by the spinal column in the adult stage of the latter.

**Nucleus** Biology: The living portion of the cell that contains chromosomes and serves as the "control center" of the cell. Physics: The central part of the atom containing protons and neutrons.

**Organismic succession** The principle that living organisms have appeared in a unique sequence—each form appearing only once in the history of life. From this principle is derived the conclusion that all rocks containing a specific fossil assemblage must be of the same relative age (*see* Fossil correlation).

**Phenotype** The physical condition produced by a particular genotype. (*see* Trait).

**Phloem** A type of vascular tissue that conducts food within the plant.

**Photosynthesis** The process in which light energy trapped by chlorophyll is used to produce sugar molecules and oxygen gas ($O_2$).

**Phylogeny** The evolutionary history of a group of organisms.

**Placenta** The connection between mother and embryo through which food and waste products are exchanged during pregnancy.

**Polygenic trait** A trait that is controlled by several pair of alleles occurring at different loci.

**Polyphyletic model** A phylogenetic model in which different groups of organisms have evolved from different ancestors.

**Population** A group of individuals that belong to a single species.

**Procaryotic cell** A cell that lacks a nuclear membrane and cytoplasmic specialization (e.g., bacteria and blue-green algae).

**Protein** A cellular constituent composed of a chain of amino acids.

**Protista** Primitive organisms with procaryotic cells.

**Race** A genetically distinct subdivision of a species population that is usually produced by geographic isolation.

**Recapitulation** The theory that stages of embryonic development of an organism repeat events of its evolutionary history.

**Recessive** A gene whose expression is suppressed by the dominant allele.

**Recombination** A new combination of genes appearing in progeny of genetically dissimilar parents. The result of random assortment of genes.

**Relative age** Determination that rock strata are older, younger, or the same age through use of such principles as superposition or fossil correlation.

**Reproductive isolation** Inability of members of two different species to interbreed successfully.

**Respiration** The release of energy from organic compounds. In the living cell this process is controlled by respiratory enzymes.

**Rhizoid** A rootlike structure that, in the bryophytes, serves as a mechanism of attachment to the substrate. Rhizoids do not have internal vascular tissue.

**Ribosome** A subcellular unit found in the cytoplasm of eucaryotic cells. It functions in protein synthesis.

**Root** A specialized structure found in vascular plants that anchors the plant in the substrate and absorbs water and dissolved minerals from the soil. Roots contain xylem and phloem.

**Seed** An organ that consists of embryo and seed coat. It may also contain a supply of stored food.

**Spatial isolation** Geographic isolation of units of a population as a result of distribution over a very large geographic area.

**Speciation** The formation of a new species.

**Species** A population of interbreeding organisms that is reproductively isolated from other species.

**Sporangium.** A structure in which spores are produced.

**Sporophyll** A leaf specialized to produce spores.

**Sporophyte** In alternation of generations: the diploid plant that produces spores by meiosis.

**Stratum** A layer of sedimentary rock (plural, strata).

**Subpopulation** An identifiable subunit of a species population.

**Superposition** The geologic principle that states that in any undisturbed series of sedimentary rock layers, the bottom-most stratum is the oldest and the topmost stratum the youngest. A relative dating technique.

**Symbiosis** An intimate, long-term relationship between members of two species.

**Tetraploid** A nucleus that contains four sets of chromosomes.

**Tracheid** An elongate cell with thickened cell wall found in xylem tissue. Tracheids function in water transport and in support.

**Trait** The physical condition produced by a particular genotype (*see* Phenotype).

**Triploid** A nucleus that contains three sets of chromosomes.

**Vertebrates** Animals with a backbone.

**Xylem** A type of vascular tissue that conducts water and dissolved minerals.

**Zygote** The fertilized egg; a diploid cell resulting from the union of sperm and egg cells.

# Supplementary Readings

AVERS, C. J., *Evolution.* New York: Harper and Row, Publishers, Inc., 1974.

BERRY, W. B., *Growth of a Prehistoric Time Scale, Based on Organic Evolution.* San Francisco, Calif.: W. H. Freeman & Company Publishers, 1968.

COLBERT, E. H., *Evolution of the Vertebrates* (2nd. ed.). New York: John Wiley and Sons, Inc., 1969.

DARWIN, C., *On The Origin of Species: A Facsimile of the First Edition.* Cambridge, Mass.: Harvard University Press, 1976.

DEBEER, G., *Charles Darwin.* Westport, Conn.: Greenwood Press, Inc., 1976.

DOBZHANSKY, T., *Mankind Evolving: The Evolution of the Human Species.* New Haven, Conn.: Yale University Press, 1962.

EICHER, D. L., *Geologic Time* (2nd. ed.). Englewood Cliffs, N.J.: Prentice-Hall, Inc., 1968.

FENTON, C. L. and M. A. FENTON, *The Fossil Book.* New York: Doubleday & Co., Inc., 1959.

GOIN, C. J. and O. B. GOIN, *Journey Onto Land.* New York: Macmillan Publishing Co., Inc., 1974.

HANSON, E. D., *Animal Diversity* (2nd ed.). Englewood Cliffs, N.J.: Prentice-Hall, Inc., 1964.

LEAKEY, R. E. and R. LEWIN, *Origins.* New York: E. P. Dutton and Co., Inc., 1977.

MOLNAR, S., *Races, Types, & Ethnic Groups: The Problem of Human Variation.* Englewood Cliffs, N.J.: Prentice-Hall, Inc., 1975.

MOORE, J. A., *Heredity and Development* (2nd ed.). New York: Oxford University Press, 1972.

MOORE, R., *Man, Time, and Fossils.* New York: Alfred A. Knopf, Inc., 1961.

OLSON, E. C. and J. A. ROBINSON, *Concepts of Evolution.* Columbus, Ohio: Charles E. Merrill Publishing Company, 1975.

PHENICE, T., *Hominid Fossils: An Illustrated Key.* Dubuque, Iowa: William C. Brown Company, Publishers, 1972.

POIRIER, F. E., *Fossil Evidence: The Human Evolutionary Journey.* St. Louis, Missouri: The C. V. Mosby Co., 1977.

ROMMER, A. S., *Vertebrate Paleontology* (3rd ed.). Chicago, Ill.: University of Chicago Press, 1966.

STEBBINS, G. L., *Processes of Organic Evolution.* Englewood Cliffs, N.J.: Prentice-Hall, Inc., 1971.

VOLPE, E. P., *Understanding Evolution.* Dubuque, Iowa: William C. Brown Company, Publishers, 1977.

WASHBURN, S. L. and R. MOORE, *Ape into Man.* Boston, Mass.: Little, Brown & Company, 1974.

YOUNG, L. B. (ed.), *Evolution of Man.* New York: Oxford University Press, 1970.

# Index

| Millions of years ago | Geologic Era | Geologic Period | Major Events<br>First appearances of some major groups of living organisms | |
|---|---|---|---|---|
| | Cenozoic | Quaternary | | |
| 2 | Cenozoic | Tertiary | Hominids<br>Anthropoids | |
| 65 | | | Primates | |
| | Mesozoic | Cretaceous | Marsupials Placentals | |
| 136 | | | | Angiosperms |
| | Mesozoic | Jurassic | Monotremes<br>Mammals Birds | |
| 190 | Mesozoic | Triassic | First dinosaurs | |
| 225 | Paleozoic | Permian | | Conifers |
| 280 | Paleozoic | Carboniferous | Reptiles | "Seed ferns"<br>Ferns |
| 345 | Paleozoic | Devonian | Amphibians<br>Bony fish Sharks | Horsetails |
| 395 | Paleozoic | Silurian | Armored fish | Club mosses<br>Psilopsids |
| 430 | Paleozoic | Ordovician | Jawless fish | |
| 500 | Paleozoic | Cambrian | "Age of Invertebrates" | |
| 570 | Precambrian | | First multicellular organisms (invertebrates)<br>First eucaryotic cells<br>First procaryotic cells (Bacteria and blue-green algae) | |
| 4500 | | | | |

Geologic Time Scale